Beavers

& Other Pond Dwellers

A TIME-LIFE TELEVISION BOOK

Editor: Eleanor Graves
Series Editor: Charles Osborne
Text Editor: Richard Oulahan
   Associate Text Editor: Bonnie Johnson
   Author: Ogden Tanner
   Assistant Editor: Peter Ainslie
   Literary Research: Ellen Schachter
   Text Research: Ann Guerin
   Copy Editors: Robert J. Myer, Greg Weed
Picture Editor: Richard O. Pollard
   Picture Research: Judith Greene
   Permissions: Cecilia Waters
Designer: Robert Clive
   Art Assistant: Carl Van Brunt
Production Coordinator: Jane L. Quinson

WILD, WILD WORLD OF ANIMALS
TELEVISION PROGRAM
Producers: Jonathan Donald and Lothar Wolff
This Time-Life Television Book is published by Time-Life Films, Inc.
Bruce L. Paisner, *President*
J. Nicoll Durrie, *Business Manager*

THE AUTHOR

OGDEN TANNER, a former senior editor of TIME-LIFE Books, writes on nature and other subjects. In addition to articles on coyotes and the impact of man on the environment, he is the author of books in the TIME-LIFE Books American Wilderness series, *New England Wilds* and *Urban Wilds* and of *Bears & Other Carnivores* in the *Wild, Wild World of Animals* series. A native New Yorker and an architectural graduate of Princeton University, he has been a feature writer for the San Francisco *Chronicle*, associate editor of *House & Home* and assistant managing editor of *Architectural Forum*.

THE CONSULTANTS

WILLIAM H. AMOS, a noted biologist, is Chairman of the Science Department at St. Andrew's School in Middletown, Delaware, and a Research Associate at the University of Delaware. He has been associated with the New York Zoological Society, the American Institute of Biological Sciences and the Marine Biological Laboratory at Woods Hole, Massachusetts. He has written a number of books, including *Life of the Seashore*, *Life of the Pond* and *The Infinite River: A Biologist's Vision of the World of Water*. A frequent contributor to *National Geographic* and *Scientific American*, Mr. Amos also has an extensive collection of photographs of marine and freshwater subjects.

SIDNEY HORENSTEIN is on the staff of the Department of Invertebrates, at the American Museum of Natural History, New York, and the Department of Geology and Geography, Hunter College. He has written many articles on natural history and has been a consultant on numerous Time-Life books. He publishes *New York City Notes on Natural History* and is Associate Editor of *Fossils Magazine*.

DOUGLAS A. LANCASTER is Director of the Cornell Laboratory of Ornithology. His research specialty is the study of the ecology and behavior of birds of the New World tropics. He has written numerous articles on the subject for scientific journals and has conducted ornithological expeditions in Central and South America.

DR. RICHARD G. ZWEIFEL is Chairman and Curator in the Department of Herpetology of the American Museum of Natural History in New York. His fields of study include the ecology and systematics of reptiles and amphibians, in particular those of America and New Guinea. Dr. Zweifel has published more than 70 scientific papers in addition to semipopular articles for magazines and encyclopedias. His memberships include the American Society of Ichthyologists and Herpetologists, the Herpetologists League, the Society for the Study of Amphibians and Reptiles, the Ecological Society of America and the Society for the Study of Evolution.

Wild, Wild World of Animals

Beavers
& Other Pond Dwellers

Based on the television series
Wild, Wild World of Animals

Published by
TIME-LIFE FILMS

ISBN 0-913948-16-0

Library of Congress Catalog Card Number: 77-82678

Printed in the United States of America.

Contents

Introduction
by Ogden Tanner

THE POND IS ONLY ABOUT 100 FEET across and no more than four feet deep, little more than a pretty, irregular widening in a tranquil stream. Through the winter it hibernates, silent except for the cracking and groaning of ice and occasional shouts of children who come to skate when the ice is good. Most of the time the only sounds are the baleful caws of crows flying overhead and the mournful soughing of the wind in the leafless trees.

On early spring mornings, however, the pond begins to take on a pulse of its own. Through the greening thickets around the water eerie mists billow up like primordial steam, to be chased into wispy tatters by a rising breeze. A warming sun breaks above the treeline, shattering the wind-rippled surface into a thousand points of dancing light. A pair of mallards scribes slow, straight lines across the water, and a flight of Canada geese splashes down noisily on the way to summer nesting grounds in the north. A red-winged blackbird clings cockily to a willow branch and proclaims its new territory with a shrill, triumphant call. As the weather warms still more, the evenings begin to buzz with the songs of insects and the high-pitched chorus of a hundred tiny, unseen frogs. Gradually the water itself seems to come to life—a simmering animal-vegetable soup of crawling, swimming, hopping, flying creatures and swaying, undulating plants. It is a watery world, a primal world in miniature, being created all over again.

The word "pond" generally denotes a small body of water, "lake" a larger body. But the fact that the terms are vague underlines their close kinship. There are millions of such freshwater catchments scattered all over the world, each supporting its own distinct community of life. They range from shallow, acre-sized New England ponds to the Soviet Union's awesome Lake Baikal, where freshwater seals play about the 12,160-square-mile surface and strange, transparent fish lurk in the darkness almost a mile below. On the barren Gobi and Sahara deserts are green-ringed oases that provide water for thirsty men and animals as well as homes for permanent populations of tiny aquatic life. In the dry regions of the American West, even salty lakes and alkaline water holes are the harsh habitats of brine shrimp, algae, brine flies and protozoans. The treeless wastes of the Arctic tundra are pitted with thousands of rocky, water-filled basins left by the past ice age, known to such creatures as mosquitoes, black flies, water fleas, lake trout, whitefish, numerous birds, including sandpipers and plover, and occasional wolves, musk-oxen, caribou, wapiti and Arctic hares.

It is in the temperate and tropical regions of the world, though, that abundant rainfall, warmth and nutrients enrich ponds and lakes with their greatest variety and volume of living organisms. In such regions the still or slowly circulating waters, warmed by the sun, teem with living things, often reaching a population density that is unmatched by any other watery habitat except, perhaps, a tidal estuary or a tropical coral reef. In and around these waters lives a staggering collection of organisms, ranging from microscopic, single-celled plants and animals to 200-pound turtles and giant mammals such as the half-ton bull moose.

Warm coastal plain pond, North Carolina

No two ponds or lakes are quite alike. Each, like the plants and animals it harbors, is living out a complex life cycle of its own. The subtle differences between ponds and lakes have led to a confusion in terminology, on which even professional limnologists, the scientists who specialize in the study of inland waters, do not always agree. Broadly speaking, it can be said that lakes are inland depressions containing standing water and that ponds are smaller than lakes, not large enough to allow major waves to gather and erode their shores and shallow enough to maintain more or less uniform water temperatures and to permit rooted plants to receive enough sunlight to grow on most of the bottom. A pond, though it may be relatively large in area, is seldom more than 12 to 15 feet deep, the maximum depth at which sunlight can penetrate sufficiently to nourish plants. A lake may be relatively small, yet, particularly in mountain regions, may reach a depth of 100 feet or more, too deep for plants to grow except near the shores.

This book concerns itself primarily with the richer life provided by the pond, even though within that focus there are marked variations. In terms of the kind of life nurtured by ponds, limnologists make three basic distinctions. Most shallow ponds in warmer regions are eutrophic, or richly nourishing, amply supplied with dissolved nutrient compounds containing nitrogen, phosphorus and calcium. Because of a heavy accumulation of decaying organic matter on the bottom, though, they may become oxygen-short by the end of summer. As a result, the animal communities they support may be inhibited and limited. Many cold, clear northern ponds are oligotrophic, or scantily nourishing, high in oxygen content but low in nutritive minerals. As a result, the animal and plant life they are able to support must be hardy and adaptable. Some ponds are called dystrophic, or faultily nourishing. The most familiar of these are bogs, where poor drainage and cool microclimates may create conditions low in both nutrients and oxygen,

Glacial lake, Brooks Range, Alaska

high in carbon dioxide and strongly acidic. Often dominated by floating, encircling mats of sphagnum moss, such swampy environments can support only their own special, limited communities of animals and plants.

Though they may seem like permanent features of the landscape, many ponds are surprisingly transient; over varying periods of time they come and go. A particular pond observed today may have been formed in a number of different ways: by the action of the last ice age, which left countless, gouged-out depressions and dams of rock to fill up with rainwater; by a landslide in a steep-walled mountain valley that dams up the river; by a meandering river that shifted its course and cut off one of its own wayward loops to leave a stranded "oxbow" of still water; or by a river that periodically floods onto the adjacent plains along its course, turning depressions into swampy meadows and cypress ponds. An increasing number of ponds today are man-made: old mill ponds origi-

nally dammed to power water wheels, farm ponds, reservoirs, park ponds and other bodies of water created to provide water supply, power, recreation or decoration.

Many ponds in northern temperate zones are the works of beavers (pages 18–31), whose skill at dam building has won the admiration of scientists and engineers. Like many natural bodies of water in wild or semiwild areas, beaver ponds not only support a wide range of smaller aquatic creatures but provide water and food that attract larger animals as well. They also help maintain a fine ecological balance by controlling runoff, erosion and floods and by keeping water tables high enough to support rich vegetation. When beavers begin to exhaust the food and building materials supplied by shrubs and trees near their pond, they may enlarge the dam and the pond or dig canals to reach distant trees—or move elsewhere to start another pond. Behind them they leave a wet place that slowly fills up with rich silt and decaying plants. Eventually it becomes meadowland, ideal for grazing or cultivating crops. Such beaver-pond legacies, if left alone long enough, may in turn give rise to a fresh growth of young trees and shrubs. The early settlers of North America probably had generations of beavers to thank for the fertile, deep-soiled bottomland they found to put to the ax and plow.

Regardless of how they were formed originally, all ponds are subject to much the same laws of succession, a term used by ecologists to describe the gradual replacement of one community of associated plants and animals by another until a "climax" or self-perpetuating community is attained. In simple terms this means that ponds, like the flora and fauna in and around them, are born, grow old and die. Generally speaking, the larger and deeper a body of water, and the farther north its location or the higher its altitude, the longer it will last. In the Arctic tundra, the permanently frozen ground locks up sources of mineral nourishment, sparse rainfall washes in little silt and short summers allow minimum plant growth, with the result that ponds may remain for thousands of years with no discernible change. Ponds in warmer, wetter, more fertile regions, on the other hand, may be overrun by luxuriant vegetation and fade away in a mere human generation or two.

For most of their lifetimes, however, ponds are free to perform their major role as prolific breeders and feeders of life. A brand-new pond, dammed by beavers or scooped out of raw earth by a bulldozer on a farm, slowly fills with water from an inlet stream and rain. The water washes in tiny plant and animal organisms and the seeds, eggs and spores of larger ones. Ducks, geese and wading birds bring in still more seeds and eggs stuck to their muddy feet and feathers. Strings of algae start to bloom in the shallows, and shoreline rushes begin to germinate from wind-borne seeds. Attracted by the growing skein of life, insects fly in to consume smaller insects, and frogs migrate from other nearby waters to eat the insects.

Before long the edge of the pond becomes fringed in a soft, ascending collar of different plants. In sheltered waters duckweed provides a home and food for still more animals, including aquatic insects, crayfish and snails, and the snakes, turtles and fish that feed on

Great Swamp Wilderness Area, New Jersey

them. Sunfish and minnows flit through the shallows after small organisms and in turn are consumed by pickerel and largemouth bass. As more and more plant and animal matter dies and drifts to the bottom, building up a rich muck and decreasing the water depth, water lilies take root farther and farther from the shore. On the protected undersides of their leaves entire communities of smaller organisms take up residence or lay their eggs: protozoans, rotifers, snails, hydras, water mites, beetles, sponges, worms.

Gradually larger animals arrive, not only to drink and seek relief from summer heat but also to take advantage of the pond's growing wealth of food: muskrats to feed on the marsh grasses and build winter lodges out of their stems; deer and moose to browse on tender aquatic shoots; raccoons to scout the shallow water for crayfish, mussels and snails. If the area is wild enough a stealthy, swift-bodied mink may slip in to prey on the muskrats, and otters may stop by to dive for fish.

In such myriad ways a complex food chain, or feeding pyramid, is established in and around the pond. At its broad base are the lowly but indispensable manufacturers of food, the millions of tiny plants that make up the pond's phytoplankton: algae, diatoms, bacteria, desmids, flagellates. Like the larger pond plants that furnish food for muskrats, ducks and turtles, these tiny food factories convert sunlight into sugar energy through photosynthesis. The phytoplankton, which may make up 90 percent or more of the pond's microscopic life, is consumed, along with large quantities of detritus, by the other 10 percent, the zooplankton, minuscule animals such as rotifers, new hatchling fish, water fleas and copepods. Small fish and insects gobble up quantities of them and in turn are eaten by larger fish, insects, crustaceans, frogs and birds. These thereupon fall prey to the largest animals at the top of the pyramid: big fish, snapping turtles, predatory mammals such as otters and mink and birds such as ospreys and other hawks. Since roughly 90 percent of the food energy consumed is lost at each step up the chain—some of it burned up as fuel simply to keep the eaters breathing, digesting and moving around, some of it added as weight to the bodies of the animals and some expelled

14   *Badwater, a saline pond in California's Death Valley*

as waste—the pyramid sharpens rapidly to a point. A thousand pounds of phytoplankton can produce only 100 pounds of zooplankton; 100 pounds of zoo-plankton can produce only about 10 pounds of small predators, which in turn can produce only one pound of snapping turtle or osprey at the top of the pyramid.

As the pond reaches maturity and approaches old age it becomes increasingly clogged with plants. Duckweed and water lilies form dense floating mats that shade out and kill the forest of underwater vege-tation, which then sinks to the bottom and thereby forms a rich rooting ground for the next succession of plants. Cattails and rushes root farther and farther out toward the narrowing circle of open water as the inlet stream deposits more and more sediment around their stems. Slowly, almost imperceptibly, as the pond becomes shallow and diminishes in area and aquatic life, it becomes a marsh. In areas where water-tolerant shrubs and trees such as alder, swamp maple or cypress can gain a foothold in the muck, the marshy pond becomes a swamp. If poor drainage and cold, sour, tea-brown water favor sphagnum and acid-tolerant plants such as leatherleaf, the aging pond becomes a bog and may eventually be taken over by cedar, black spruce, larch or balsam fir. Whatever route it follows, the dying pond becomes so full of mud and vegetation that fewer and fewer aquatic spe-cies can live in it. Decay uses up more and more of the dwindling oxygen, slowly driving the survivors out. In the remaining pools the only fish are hardy species such as mud minnows, bullheads and shiners.

Often man steps in abruptly to stop this long and intricate process. Under the pressures of modern life, countless numbers of apparently uneconomic marshes, swamps and bogs have been buried under town garbage dumps or filled in and paved over to serve such purposes as pizza parlors and supermarket parking lots, with little or no thought of the ecological consequences. It is only in recent decades that the true functions of freshwater wetlands have become widely recognized—as way stations and nesting areas for wildfowl, as natural regulators against floods and filters of stream flow and water supplies, as spawning grounds and nurseries for whole pyramids of living things.

Pond becoming a marsh, Yellowstone Park, Wyoming

Beavers

As master engineer, lumberjack and pond builder of the animal world, the beaver has long been highly esteemed. Not only have its sheer industry and dogged perseverance inspired such terms as "busy as a beaver" and "eager beaver," its skills have also provided men with helpful hints on everything from swimming techniques to natural fortification, dam building and flood control.

The known ancestry of dam-building beavers goes back five million years. One early beaver grew to be almost as large as a black bear. Modern beavers, native to the Northern Hemisphere in both the New and Old Worlds, are still among the largest rodents, or gnawing animals. An adult weighs 40 to 60 pounds and measures four feet long, though some fat old beavers have tipped the scales at over 100 pounds. About a foot of the length consists of a unique, flattened, paddle-shaped tail six or seven inches wide.

Beavers swim expertly at speeds of up to six miles per hour by pushing against the water with their strong, webbed hind feet. The webbed feet of the beaver doubtless helped inspire the inventor of rubber swim fins—though at least one naturalist, using large fins on his feet, tried to race a beaver underwater and found that he could not begin to keep up. The beaver's smaller forepaws are not webbed but have long, powerful digging nails. When swimming, the animal balls its forepaws into little fists and uses them to fend off floating objects or to carry loads of twigs and mud close to its chest and chin.

To enable it to swim and work in icy water, a beaver has a heavy layer of fat under its skin and a dense, rich coat of fur that it keeps in prime condition by frequent grooming. Two special split nails on each hind foot serve as combs. To waterproof its coat, the beaver applies an oily secretion called castoreum from scent glands located near its anus.

Beavers living by a large river or lake may simply burrow into the side of the bank to make a den, but their most spectacular activities involve damming a stream to create a pond where they can build a protected island lodge. Beavers are not always precise surveyors. Occasionally they will choose a site that requires a dam 100 feet long, when they could have picked a place not far up- or downstream that would have enabled them to construct a much shorter barricade. Once they have chosen a location, however, almost nothing can stop beavers from getting the dam built and carefully keeping it repaired. The principal materials for both food and construction are alder, willow and aspen trees, which often grow in dense thickets along the sides of streams. Using powerful jaw muscles and long gnawing teeth—which have a front layer of hard orange enamel that their owner can sharpen to a chisel edge—a beaver cuts rapidly, slicing chips up to six inches long out of soft woods; it can drop an alder bush in eight or 10 bites and fell a willow tree five inches in diameter in three minutes.

Downed trees and bushes are cut into portable lengths, which the beavers drag with their teeth to the dam site, jamming the sharp ends of some into the ground to anchor them and pointing others into the current so they will not be swept away. In fast-moving streams the shape of the dam may be bowed upstream to counteract the water pressure, like many man-made dams. Gradually a pile of brush rises evenly along the length of the dam, sometimes weighted down and secured by rocks weighing 30 or 40 pounds— almost as heavy as the beavers themselves. Mud, leaves and other debris plastered along the upstream side caulk the dam and make it watertight. To sluice off flood water that might harm the dam, a temporary spillway may be constructed at one end. Smaller subsidiary dams are sometimes built downstream to raise the water level on the lower face of the main dam and relieve the pressure, or upstream to flood a greater area and widen accessibility to more trees for building and food. Dams vary widely in their design, depending on the topography and the number of years they have been built and maintained. One found in a steep-sided Wyoming valley was 18 feet high but only 30 feet across. The greatest beaver dam ever recorded, built near Berlin, New Hampshire, was 4,000 feet long, backing up a lake in which, over the years, no fewer than 40 beaver lodges had been built.

Most of this prodigious construction is done in the hours after sundown. Beavers have proven relatively easy to trap, even at night, and the 19th-century craze for lustrous beaver-felt hats and furs nearly caused the extermination of the beaver population before regulatory laws were passed and fashions changed. Today, under protection, the beaver has made such a comeback that in some areas it has become a nuisance, flooding farmers' fields, valuable timberland and busy highways, and animals frequently have to be trapped alive and moved to more remote regions where they can continue their pond building in peace.

Beaver constructing dam, Wyoming

Beginning with the leaves, a beaver nibbles on a willow branch (above). When the foliage is eaten, the beaver will consume the bark, turning the branch in its paws as humans turn an ear of corn, and then frugally use the remaining stripped wood as building material for a dam.

Forepaws held close to its chest, a beaver (right) swims underwater near its Wyoming dam. In winter, air exhaled underwater is trapped beneath the ice, reoxygenated by the water and breathed by beavers again. Even without surfacing for air beavers can swim more than a half mile underwater.

Complete Beaver

The beaver is a marvel of amphibious design. The moment one dives under water, its hatches, like a submarine's, are automatically battened. Valvelike ears and nostrils close off, and its mouth is sealed by special skin flaps that leave its large front teeth exposed for use. Clear membranes slide into place to cover its eyes like contact lenses, protecting the eyeballs from floating debris. These adaptations, plus the beaver's webbed hind feet, which quicken its swimming speed, its handlike forepaws for grasping objects and pushing debris aside, and its broad, flat tail—which functions as a rudder underwater and as a prop on land when the beaver sits upright (left)—equip the creature handsomely for its underwater chores. With its oversized lungs and liver to accommodate ample reserves of air and oxygenated blood, a beaver can stay submerged for as long as 15 minutes.

Friend or Foe?

More than almost any other creature except man, the beaver excels at altering its own domain. And with the number of beavers increasing in recent years, a controversy has developed over the right of the resourceful creature to the land it alters. Beaver ponds flood croplands, commercial timber stands and highways, the argument goes, and beaver dams occasionally block the upstream path of spawning salmon. Environmentalists counter by pointing out that the dams help reduce erosion and raise the water table of the surrounding countryside, creating a desirable habitat for wildlife of all sorts. Insects lay eggs in beaver ponds, providing essential food in the form of aquatic larvae for fish. Waterfowl, muskrats, otters and minks move in; moose and deer are drawn to the ponds to feed and drink. Such benefits to wildlife, the environmentalists contend, far outweigh any harm that beavers may do.

Alerted to a break in the dam by the sound of rushing water, a beaver arrives to make repairs with a branch. After several trips carrying sticks, the beaver seals the breach with mud and grass, scooped up and held between its forelegs and chest.

Like stairsteps down a stream, beaver dams created this series of ponds (left) in the Allagash-Katahdin area of Maine. Beavers occupy such sites for decades—even centuries—building mammoth dams like the one in New Hampshire measuring three quarters of a mile.

Gathering food for its family (above), a beaver swims homeward with a poplar branch in tow. The branch will be anchored underwater in the mud of the riverbed as part of the family's winter provender. Naturalists observed one such cache that formed a 60-foot-long, five-foot-high wall of willow branches.

Sniffing the air for predators before leaving its watery world (left), a beaver with the wet look emerges from its pond. Much less agile on land than in water, beavers are especially wary when they are ashore, interrupting their chores frequently to look around. The edge of the pond is a favored dining spot, and beavers tow their freshly cut branches there, secure in the knowledge that safety is only a dive away.

Standing on a pile of wood chips next to a half-cut cottonwood tree, a beaver (right) scans its surroundings for signs of danger. Beavers circle a tree as they cut away chips, so that the tree usually falls on the side with the heaviest foliage. Often this is toward a nearby pond, where abundant sunlight has caused thicker growth. The weathered look of the cuts in this cottonwood indicates that beavers may have abandoned it without felling it—a not uncommon occurrence but one that naturalists cannot explain.

Young and Eager

Beaver babies are born in the spring—and by no coincidence. Fresh food is plentiful at that time of year, and beaver mothers, busy in summer and fall with dam building and with gathering and storing up the winter's food supply, have ample time in spring to nurture the infants.

The babies, or kits, usually arrive in litters of two to five. Born with eyes half open, the furry kits are able to see immediately, and they often take to the water inside their lodge before they are a half hour old. Learning to swim and dive (right) is a bigger step. The kits are generally too buoyant at birth for diving; in any case, they usually stay close to their mother in the lodge for the first few weeks, nursing frequently (below) and gaining considerable weight. Once they start growing, beavers never stop. Adults range from 30 to 70 pounds, but if they live long enough—23 years is the longest life-span on record—they may reach 100 pounds.

An accomplished swimmer after one week of life, a beaver kit climbs on its mother's back for a rest. On land, females often carry their young on their flat, broad tails, and some have startled observers by walking upright, holding the kits in their forepaws in a very human manner.

Beaver kits, hungry and in a hurry, clamber over one another to get at one of their mother's four nipples. Females frequently sit on hind legs, nursing their infants in an upright position. At age two, the kits discover that they are no longer welcome in the family circle and leave to start a new home elsewhere.

In the living chamber of a beaver lodge, a kit awaits its mother's return. Wood chips on the floor absorb excess moisture, and a built-in vent provides fresh air. Adult beavers mate for life and usually share a pond with a colony that rarely contains more than 12 animals occupying two or more lodges.

Mother and kit share a meal outside the lodge. With incisors visible at birth, kits are soon consuming bark, leaves, herbs and water plants, along with their mother's milk. Though adept in the water, tired young beavers often hitch a ride on their mother's back (overleaf).

Raccoons and Otters

Sharing the pond and its environs with the beaver is a variegated company of smaller mammals. They include such familiar characters as the deft raccoon and the comical otter, as well as such tiny exotics as a featherweight shrew that walks on water and an amphibious mole with a nose shaped like an asterisk. Collectively, they are among the liveliest and most attractive of the wetlands creatures.

In contrast to the sobersided, industrious beaver, the otter is the playboy of the pond. Whatever it is doing, whether swimming, hunting, traveling overland or just relaxing, this sleek, graceful animal seems compelled to make a game of it, sometimes with hilariously clownlike results. The fun-loving river or "land" otter—so called to distinguish it from its saltwater cousin the sea otter—is a member of the weasel family, which is generally noted for its aggressiveness and vicious temper. Though an otter will put up a fierce fight if cornered, it is the pond's most happy-go-lucky creature if left unmolested.

Otters feed on fish, crayfish, frogs, salamanders, eels, insects, worms, turtles, snakes, snails, freshwater clams and mussels and an occasional muskrat or small beaver, as well as on all kinds of berries and roots. Their food sources are so abundant and otters are so adept in exploiting them that the animals feel no urgency about where their next meal might be. Even in winter, when most other animals are either hibernating or desperately looking for food, otters are cavorting over the snow with an abundance of food readily available under the ice of the pond.

Life is tolerable, if not easy, for other, smaller mammals that live in and around the pond. One of the commonest is the muskrat, a North American native that has been introduced into Europe and Asia and is now widely distributed in those regions. A small, ratlike version of the beaver and a good swimmer weighing only about three pounds, it builds protective lodges and is heavily trapped for its fur.

Other small mammals seen around ponds include swamp rabbits, rice rats, bog lemmings and the star-nosed mole, an active swimmer and tunneler that often forages along muddy bottoms for insects and worms. The smallest of all aquatic mammals are the water shrews of both the Old and New Worlds, which are usually no more than three inches long and are two to three times the weight of a 25-cent piece. What the water shrew lacks in size, however, it makes up in feverish activity to satisfy its appetite, which is so demanding that a shrew must eat several times its own weight in food every day. Though it moves constantly in and out of the water, a shrew never actually seems to get wet. Its fur is so fine that the hairs not only repel water but retain a silvery film of air bubbles around its body when it dives to the bottom in search of food. Because of its light weight, its speed and the fringes of stiff hairs on its feet, the little creature can also literally walk on water, using the surface tension to support it for short distances. Shrews must support a metabolism that can exceed 1,200 heartbeats a minute. Dashing wildly about in search of food, they live fast and die young, burning out at the ripe old age of a year and a half.

While otters, beavers and muskrats are much more aquatic than the raccoon, that bright, restless fellow with the black "bandit's" mask across its eyes and the unmistakable rings around its tail is nonetheless a regular frequenter of ponds and streams. Raccoons are natives of North America and usually live close to fresh water, where they do much of their foraging. They swim quite competently when the occasion demands.

The original Algonquin Indian name from which "raccoon" derives, arakun, means "he who scratches with his hands." While a raccoon often appears to be meticulously washing a crayfish or snail before dining, its reputation for fastidious cleanliness is undeserved. It is more likely to be feeling its food for texture and size. A raccoon is insatiably curious, handling everything with its long, sharp-clawed "fingers." Despite the fact that it has no opposable thumbs, its digits are so agile and sensitive that they can be used alone to locate food, even in pitch dark. The appearance of washing food after seizing it may simply reflect the fact that the raccoon's sense of touch is heightened when its fingers are wet.

Raccoons are among the most intelligent and adaptable animals, and despite long persecution as targets for "coon hunts" and furs for raccoon coats, they have managed to survive and proliferate in an increasingly urban world. Though they prefer to den in hollow trees near fresh water, they will take up quarters anywhere the opportunity presents itself, including woodchuck burrows, old drainpipes and even unused attic space. They will eat almost anything, and they are well aware of easy pickings provided by human neighbors. More than one angry farmer has found his hen coop, cornfield or melon patch decimated by night-roving bandits, and even tightly lidded garbage cans usually present no problem for their light-fingered burglar's touch.

Masked "Bandit"

Although they have been hunted extensively for their meat and their fur, raccoons have continued to survive and thrive and remain among the most familiar mammals of North America. Ranging from southern Canada across the United States into Central America, these creatures are easily recognized by their five to 10 black tail rings and the mask of black fur across their eyes.

Raccoons are solitary, nocturnal animals. They frequent wooded areas close to fresh water. In this habitat the raccoon has an omnivorous appetite for such pond creatures as insects, fish and frogs, as well as seasonal provender such as fruits, nuts, seeds, acorns and eggs.

In the southern part of their range raccoons are active all year long. But in the north, where the harsh winters bring severe winds, deep snow and temperatures below 25° F., the raccoon's activity drops off sharply. It may remain for weeks in its burrow without venturing out for food. During this period the raccoon sleeps; but since it does not experience a markedly lowered heart rate, metabolic rate and body temperature, it is not in a true state of hibernation.

Two raccoons (above and at left) scout the edges of a pond looking for likely prey. Although they are accomplished swimmers, raccoons usually avoid going into deep water. They track down their food using highly developed senses of touch and smell, grabbing prey skillfully with the long fingers on their front paws.

The Durable Muskrat

The muskrat is custom-built for life in the water. Its partially webbed hind feet are rimmed with a row of stiff, stubby hairs called swimming fringes that give the muskrat extra propulsion through the water. The muskrat's naked, scaly tail is flattened laterally and serves as a rudder, while folds of skin seal off the inner ear when the muskrat travels underwater.

Muskrats inhabit lakes, ponds, rivers, streams and fresh- and saltwater marshes, where they use cattails and other aquatic vegetation to construct mounds that may measure as much as four feet high and six feet in diameter. The muskrat's nest is tucked in the center of such a pyramid. Most young are born from November to April after a gestation period of between 22 to 25 days. Litters range in size from one to 11 young that are weaned after about one month. Females breed again while they are nursing, resulting in several litters produced each year and accounting for the muskrat's abundance in inland waterways.

Frightened by an intruder, a soaking-wet muskrat (above) bares its teeth in a menacing threat display. If this ploy fails to dissuade an enemy, the muskrat will head for the water (left), where its swimming skills and ability to stay submerged for as long as 12 minutes usually ensure its escape.

Where the Water Lilies Grow

by R.D. Lawrence

The lowly muskrat is one of the least appreciated of pond animals. Its lodges and canals are extraordinary aquatic constructions, yet they pale in comparison to the spectacular works of the beaver. Its pelt, while coveted by trappers, is not as highly esteemed as that of the mink or otter. It lacks the personality and appeal of a raccoon. Yet a muskrat has a keen instinct for survival and a dogged courage that few creatures its size can match. R. D. Lawrence, an author who lives in the wilderness of Canada, describes the confrontation of a desperate muskrat with a fox five times its size in the following excerpt from his account of life around a backwoods lake, Where the Water Lilies Grow.

Late one afternoon in February I was returning to the cabin after a long tramp through the forest when movement attracted me to the island of the beaver. I was perhaps a quarter of a mile away from it and I was not at first able to interpret the two dark, moving objects that drew my attention. The wind was blowing towards me and whatever the creatures were, they were so busy with their own affairs that they had not seen me. One was quite small, the other, larger, looked about the right size and shape to be a fox.

I moved off the ice and sought the cover of the stunted trees and shrubs at the lake's edge as I crept closer to the scene; and soon I was able to recognize the two creatures. I had been right about the fox. The other was a muskrat. They were engaged in a fierce fight.

Muskrats, though small (not counting their long scaly tails, they measure about sixteen inches in length and weigh between two and three pounds) are of stout heart. With their long, sharp chisel teeth and their strong, agile back legs, they can put up fierce resistance, and the one that was now being attacked by the red fox was defending himself well. The fox, five times the size and weight of the rat, was showing great respect for its intended victim. Each was absorbed in the other and still unaware of my presence.

The fox held his body low and continuously circled the rat, now and again slithering in quickly, seeking to get a death-grip on the back of the creature's neck, while the rat constantly swivelled its short, fat body upon its long back legs, always facing the fox. It held its short forelegs pressed tightly against its chest as it stood upright, using its tail for balance, and each time the fox crept close it jumped into the air and slashed at its attacker with its teeth, the while emitting a short, shrill cry.

I had no means of knowing how long these two had been cutting and parrying, but the snow was scarred all around them and I guessed the fight had been going on for some time. Now I was close enough to see both clearly, and I could hear the rat's little squeals as it jumped at the fox, which had evidently come close enough to its prey to receive at least two bites on its muzzle, and these were now bleeding freely.

If I can avoid doing so I do not interfere with the creatures of the wilderness and now I remained still, feeling man's pity for the besieged rat that was so close to providing a meal for the fox, yet fascinated by this primitive battle for survival; and I regretted that the light was not good enough for the taking of photographs.

Again and again the fox tried for the death-hold and each time the valiant little rat foiled his enemy's move, attacking constantly, never once giving an inch of ground, and always crying his short bursts of rage. I began to wonder how long these two would keep this up and I found myself speculating on the odds. If the fox was patient enough it must, in the end, defeat the rat, for it was bigger and stronger and of more endurance, and although the rat could inflict some painful punishment on the fox, yet the bites were not dangerously severe, while the fox needed but one opening and it would kill the rat with a bite. Evidently the muskrat had been surprised on its way to or from its den and it had been trapped in the open, unable to scurry down under the ice. Now it could only stand

and fight, for there was nowhere to hide from the teeth of the fox.

Suddenly the fox changed his tactics. Instead of making his quick rushes at the rat he backed off and sat on his haunches, his quick eyes fixed upon the muskrat, his tongue lolling and jerking out of his open mouth, his fangs messages of death for his prey. The rat stood his ground. Raised on his back haunches, his little arms tucked tightly against his chest, he returned the fox's stare, a silent, small statue etched in brown fur against the trampled snow. Each remained immobile and the tactics of the fox were now clear: the rat could not turn to run without quickly being seized and killed, and yet it had so far withstood the fox's direct attacks and had even inflicted some damage on his tormentor; so, the fox resorted to psychological warfare, hoping that the persistent stare and the sight of those gleaming teeth would eventually panic the rat, make it turn and run, and thus offer the fox an easy victory. But that rat would not be panicked.

For almost ten minutes the two remained staring at each other and it was the fox who broke first. Springing to his feet he made a fast rush at the rat, impatience clear in his every move. The rat jumped and squeaked and bit the fox on the nose. The fox backed away, hesitated and at last turned away. He travelled about five yards, stopped and looked back at the rat, who remained on his haunches ready for more trouble, then the fox loped away in obvious disgust. The small rat remained still for another five minutes, then he dropped to all fours and turned in a direction opposite to that which the fox had taken. In another few minutes he had disappeared. This time the fox had lost . . . perhaps another day the fox would win, and the muskrat would die. This is the way of the wilderness.

Aquatic Quartet

Among the mammals that live near the pond are the quartet seen on these pages. Two of them, the capybara (above) and the rice rat (right), are rodents, while the others, the mink (opposite, above) and the striped skunk (opposite, below), belong to the family Mustelidae, which includes some of the most highly prized fur-bearing animals.

The South American capybara is the largest living rodent, with adults often reaching a length of close to five feet and weighing over 100 pounds. A close but much smaller relative is the rice rat. Relentlessly exterminated as a pest because of the damage it causes in rice fields, the nine- to 11-inch rodent feeds on grasses, seeds, fruits and fish. Plants are a mainstay of the mink's omnivorous diet, which is why these lustrously furred creatures are frequently found along the shorelines of lakes and streams. Like the mink, the striped skunk is a good swimmer, although, unlike its cousin, it generally avoids going into the water.

A rice rat pokes through grass and weeds (above) searching for a morsel to eat. Rice rats also use such shoreline vegetation to construct feeding platforms near their nests, which are woven from aquatic grass and situated above the flood level.

Standing belly-deep in the water, two capybaras (left) pause in the midst of a feast of aquatic plants and grasses. Capybaras are social animals that live peaceably in family units or in colonies of as many as 20. When undisturbed they are active throughout the day and evening, but in the parts of their range where they are hunted, capybaras have become nocturnal animals.

Baring its teeth threateningly, a mink zealously guards a rat it has just caught. Rodents make up a major part of the mink's diet and are killed in typical mustelid fashion, with a bite on the back of the neck. Minks are primarily nocturnal animals that seek shelter during the day in natural cavities and depressions formed in stream banks and in rocky crevices.

The slow-moving skunk is not a bird hunter, but birds' eggs are another matter. The striped skunk at left has just come upon a nest abandoned by a duck that fled when the skunk appeared.

41

Shrews and Voles

Although they belong to completely different families within the mammalian order, shrews and voles are often confused because they appear to be so much alike. Both are small, skittish animals with long snouts, short legs, skinny tails and pelages that range in color from gray to brown to black. Voles usually prefer moist habitats such as the banks of lakes and streams, where they often dig burrows and build runways through the grass, but shrews are more varied in their habitat preference. Although both are terrestrial and spend most of their time, day and night, scurrying through the low vegetation bordering the water, voles are accomplished swimmers and divers as well. Voles, such as the water vole (opposite) and the meadow vole (below, right), are rodents. Shrews, such as the northern water shrew (right) and the wandering shrew (below), are insectivores whose large appetites for insects and insect larvae make them welcome visitors to agricultural areas.

Shrews subsist primarily on reeds and grasses as well as on various invertebrates, such as the worm captured by the northern water shrew (above). Voles, on the other hand, are strictly vegetarian. In a 24-hour period a meadow vole (below) may consume its own weight in leaves, roots and seeds.

A wandering shrew hovers protectively over her newly born litter of young. Born naked and blind after a gestation period of 17 to 28 days, the young are weaned after two to four weeks. Shrews are nervous little creatures with a brief life expectancy, and when disturbed their hearts may register as many as 1,200 beats a minute.

River Ruffians

The largest of the mustelids are members of the otter family. River otters grow to almost three feet in length and weigh over 30 pounds. Found in all types of inland bodies of water throughout much of the world except Australia, river otters are superbly suited for an aquatic existence. Their rounded heads, short, thick necks, cylindrical bodies, thick tails and webbed feet make movement through the water virtually effortless. Muscles in their ears and nostrils batten down and waterproof these organs when the otters are submerged. Otters are active throughout the day and night. When they are not occupied hunting for fish, frogs, birds, rodents or rabbits, they spend their time playfully sliding down mud- or snowbanks or cavorting in the water. For most of the year otters are peaceable animals, but during the breeding season fights often erupt between competing males. The gestation period varies with the species; the size of the litter is usually two or three. The young are nursed for about four months and remain with their mother until they are about eight months old, when they strike out on their own.

Disturbed by the sound of an intruder, a group of otters (above) ceases fishing and looks around alertly for what may be a danger to them. Otters often hunt together and when fishing will work cooperatively to corral a shoal of fish into an inlet where the chances of trapping them are much improved.

A newly caught fish dangling from its mouth, an otter (left) paddles to shore. Large catches such as this one are dealt with on land, while smaller prey is consumed in the water as the otter floats on its back. If there are many fish available, an otter will catch all it can and take them to shore before eating any of them.

Standing erect and attentive, an otter (opposite) sniffs the air for signs of potential prey or possible danger. In areas where they are not disturbed, otters come out freely during the day to hunt or sun themselves on rocks. In places where they are hunted for their velvety fur, however, they have become wary and are primarily nocturnal.

Pond Fish

The fish that adapt to life in ponds and lakes are a hardy breed. They need less oxygen than their freshwater cousins that live primarily in streams and rivers, where fast-moving water is constantly mixed with air. They can stand much greater temperature fluctuations in the still, shallow pond water, which warms up in summer to temperatures that would be intolerable to riverine fish. And they can survive nearer to the bottom in stagnant, even polluted water. For these reasons ponds tend to harbor fewer finicky game fish such as brown or rainbow trout and more "warm water" fish such as sunfish, bullheads, pickerel and perch. Nevertheless, many ponds have large and varied populations of finned inhabitants, including some fierce predators and sporting game fish.

Among the most familiar pond fish are the many kinds of minnows, the silvery species often seen flashing through the water in schools. Not all minnows are small, though they are often thought of as invariably tiny and are mistakenly lumped with any immature fish. Members of the family Cyprinida, some minnows never grow more than an inch or two long, but others, such as squawfish and carp,

may measure as much as four feet and weigh 50 pounds when they reach full maturity. Among the most numerous are the chubs and shiners, including the three-inch redfin shiner, the 10-inch creek chub and the golden shiner, which may grow over a foot long.

The largest true minnow is the carp, imported from Europe in the 19th century as a food fish and now abundant in warmer waters all over Canada, Mexico and the United States. Carp eat almost everything, from insects and fish eggs to small fish, and their habit of uprooting aquatic vegetation, both for the tender roots and for the prey hiding among them, has earned them a bad reputation among some conservationists and fishermen for muddying the waters and destroying the spawning grounds of other fish that are valued as game.

Often mistaken for larger members of the minnow family are various suckers, quiet, adaptable species that dredge up small mollusks, larvae, worms and plant matter from the mud with their down-pointing lips. Better known among pond fishermen are the sunfish, which include small, rounded, flat-bodied and often colorful species such as the

Paddlefish

pumpkinseed and orange-spotted "sunnies," crappies, rock bass, warmouth and those feisty game fish, the small- and largemouthed bass. While the smallmouth is generally found only in the cool waters of fast-flowing streams and rocky northern lakes, the largemouth can tolerate shallow, weedy waters with temperatures of 80° F. or more.

The giants among pond and lake fish are lake trout and channel catfish, which sometimes reach formidable proportions in larger bodies of water. Among the most ferocious of the larger fish are members of the pike family, ranging from the relatively small (12-inch) grass pickerel to the huge northern pike and the muskellunge of deep, cold lakes that may run well over 50 pounds. All of them— including the grass pickerel, chain pickerel and redfin, which are most commonly seen in smaller bodies of fresh water—are solitary marauders, the lone wolves of the pond. Slender and arrow-shaped like barracudas, with the latter's big jaws and deadly teeth, pikes and their cousins hide in the thickets of underwater plant stems, ready to defend their individual territories against all comers—or to dart forward from ambush with blinding speed to slaughter other fish, frogs, insects, snakes, ducklings and muskrats. The naturalist Henry David Thoreau described the pickerel as "a solemn, stately, ruminant fish, lurking under the shadow of a pad at noon, with still, circumspect, voracious eye, motionless as a jewel set in water, or moving slowly along to take up its position, darting from time to time at such unlucky fish or frog or insect as comes within its range, and swallowing it with a gulp."

Probably the strangest of all these fish, however, are the paddlefish and bowfins. Both are strong, tenacious fighters when hooked. The bowfin, which lives in sluggish waters throughout eastern North America, is among the most predatory of freshwater fish. It grows to two or three feet long and a weight of 15 pounds and is easily recognized by its cylindrical shape and long, level dorsal fin. The thicker-set paddlefish takes its name from its long, flattened snout, which it uses to scoop up plankton as it swims along. Specimens weighing 30 to 60 pounds are not uncommon, and one Brobdingnagian trophy taken from Lake Tippecanoe in Indiana measured six feet long; four feet around and weighed 150 pounds.

47

Sporting Fish

The most stagnant woodland pool is likely to contain aggressive game fish that can tolerate low oxygen content and high water temperatures that would kill most other fish. They include members of the pike family (opposite, below) and the ubiquitous sunfish (below), which, in 30 different species, is found all over the United States except in the highest reaches of the Rocky Mountains and has also been successfully introduced into European waters. One species of sunfish, the largemouthed bass, is popular among anglers as a sporting pond fish. A close relative, the smallmouthed bass, has a lower tolerance for the sluggish water of ponds and is more often found in cool, well-oxygenated streams and lakes. Because of their value as food fish, both types of bass, as well as the other fish on these pages, are frequently raised in breeding pools and marketed commercially.

The brown bullhead (above) is a frequent target for fishermen in the southern United States waters it inhabits. It has the characteristic barbels, or "whiskers," of the catfish family, which it uses as sensors to locate any kind of food, animal or vegetable, along the muddy bottom.

The grass pickerel (right) is a member of the pike family, the most ferocious predators among pond fish. Lurking under the cover of water plants, pickerels dart out with the speed of lightning to attack almost anything that moves, including smaller pickerels.

The Compleat Angler

by Izaak Walton

CHAPTER IX

Observations of the Carp, with Directions how to fish for him

Izaak Walton was a prosperous London draper who wrote a book of pious biographies of English clergymen. His greatest, indeed his only, claim to fame, however, is as the author of The Compleat Angler, *one of the earliest (1653) natural histories ever written and still, after more than three centuries, holy writ for any prospective fisherman. The book is a mix of scientific information, such as the description of a carp (below), a celebration of the English countryside and a paean to the joy of fishing.*

The Carp is the queen of rivers; a stately, a good, and a very subtle fish; that was not at first bred, nor hath been long in England, but is now naturalized. It is said, they were brought hither by one Mr. Mascal, a gentleman that then lived at Plumsted in Sussex, a county that abounds more with this fish than any in this nation.

You may remember that I told you Gesner says there are no Pikes in Spain; and doubtless there was a time, about a hundred or a few more years ago, when there were no Carps in England, as may seem to be affirmed by Sir Richard Baker, in whose Chronicle you may find these verses:

> Hops and turkeys, carps and beer,
> Came into England all in a year.

And doubtless, as of sea-fish the Herring dies soonest out of water, and of fresh-water fish the Trout, so, except the Eel, the Carp endures most hardness, and lives longest out of its own proper element; and, therefore, the report of the Carp's being brought out of a foreign country into this nation is the more probable.

Carps and Loaches are observed to breed several months in one year, which Pikes and most other fish do not; and this is partly proved by tame and wild rabbits; as also by some ducks, which will lay eggs nine of the twelve months; and yet there be other ducks that lay not longer than about one month. And it is the rather to be believed, because you shall scarce or never take a male Carp without a melt, or a female without a roe or spawn, and for the most part very much, and especially all the summer season; and it is observed, that they breed more naturally in ponds than in running waters, if they breed there at all; and that those that live in rivers are taken by men of the best palates to be much the better meat.

And it is observed that in some ponds Carps will not breed, especially in cold ponds; but where they will breed, they breed innumerably. Aristotle and Pliny say, six times in seven, if there be no Pikes nor Perch to devour their spawn when it is cast upon grass or flags, or weeds, where it lies ten or twelve days before it be enlivened.

The Carp if he have water-room and good feed, will grow to a very great bigness and length; I have heard, to be much above a yard long. It is said by Jovius, who hath writ of fishes, that in the lake Lurian in Italy, Carps have thriven to be more than fifty pounds weight: which is the more probable, for as the bear is conceived and born suddenly, and being born is but short lived; so, on the contrary, the elephant is said to be two years in his dam's belly, some think he is ten years in it, and being born, grows in bigness twenty years: and it is observed too, that he lives to the age of a hundred years. And 'tis also observed, that the crocodile is very long-lived; and more than that, that all that long life he thrives in bigness; and so I think some Carps do, especially in some places, though I never saw one above twenty-three inches, which was a great and goodly fish; but have been assured there are of a far greater size, and in England too.

Now, as the increase of Carps is wonderful for their number, so there is not a reason found out, I think, by any, why they should breed in some ponds, and not in others, of the same nature for soil and all other circumstances. And as their breeding, so are their decays also very mysterious: I have both read it, and been told by a gentleman of tried honesty, that he has known sixty or more large Carps put into several ponds near to a house, where, by reason of the stakes in the ponds, and the owner's constant being near to them, it was impossible they should be stole away from him; and that when he has, after three or four years, emptied the pond, and expected an increase from them by breeding young ones (for that they might do so he had, as the rule is, put in three melters for one spawner), he has, I say, after three or four years, found neither a young nor old

Carp remaining. And the like I have known of one that had almost watched the pond, and, at a like distance of time, at the fishing of a pond, found, of seventy or eighty large Carps, not above five or six: and that he had forborne longer to fish the said pond, but that he saw, in a hot day in summer, a large Carp swim near the top of the water with a frog upon his head; and that he, upon that occasion, caused his pond to be let dry: and I say, of seventy or eighty Carps, only found five or six in the said pond, and those very sick and lean, and with every one a frog sticking so fast on the head of the said Carps, that the frog would not be got off without extreme force or killing. And the gentleman that did affirm this to me, told me he saw it; and did declare his belief to be, and I also believe the same, that he thought the other Carps, that were so strangely lost, were so killed by the frogs, and then devoured.

And a person of honour, now living in Worcestershire,[1] assured me he had seen a necklace, or collar of tadpoles, hang like a chain or necklace of beads about a Pike's neck, and to kill him: Whether it were for meat or malice must be, to me, a question.

[1]Mr. Fr. Ru.

But I am fallen into this discourse by accident; of which I might say more, but it has proved longer than I intended, and possibly may not to you be considerable: I shall therefore give you three or four more short observations of the Carp, and then fall upon some directions how you shall fish for him.

The age of Carps is by Sir Francis Bacon in his 'History of Life and Death,' observed to be but ten years; yet others think they live longer. Gesner says, a Carp has been known to live in the Palatine above a hundred years. But most conclude, that, contrary to the Pike or Luce, all Carps are

the better for age and bigness. The tongues of Carps are noted to be choice and costly meat, especially to them that buy them: but Gesner says, Carps have no tongue like other fish, but a piece of fleshlike fish in their mouth like to a tongue, and should be called a palate: but it is certain it is choicely good, and that the Carp is to be reckoned amongst those leather-mouthed fish which, I told you, have their teeth in their throat; and for that reason he is very seldom lost by breaking his hold, if your hook be once stuck into his chaps.

I told you that Sir Francis Bacon thinks that the Carp lives but ten years, but Janus Dubravius has writ a book 'Of Fish and Fish-ponds' in which he says, that Carps begin to spawn at the age of three years, and continue to do so till thirty: he says also, that in the time of their breed-ing, which is in summer, when the sun hath warmed both the earth and water, and so apted them also for generation, that then three or four male Carps will follow a female; and that then, she putting on a seeming coyness, they force her through weeds and flags, where she lets fall her eggs or spawn, which sticks fast to the weeds; and then they let fall their melt upon it, and so it becomes in a short time to be a living fish: and, as I told you, it is thought that the Carp does this several months in the year; and most believe, that most fish breed after this manner, except the Eel. And it has been observed, that when the spawner has weakened herself by doing that natural office, that two or three melters have helped her from off the weeds, by bearing her up on both sides, and guarding her into the deep. And you may note, that though this may seem a curiosity not worth observing, yet others have judged it worth their time and costs to make glass hives, and order them in such a manner as to see how bees have bred and made their honeycombs, and how they have obeyed their king, and governed their common wealth. But it is thought that all Carps are not bred by generation; but that some breed other ways, as some Pikes do.

Time of the Trout

Brook trout, originally native only to cold, crystalline mountain lakes and streams of the northeastern United States and Canada, have been successfully introduced to similar waterways in the Western states and Europe. In the autumn the brook trout school together in nuptial pools to spawn (below). With her tail the female scoops out a depression in the rocky bottom (right) and deposits her golden roe. Within a matter of minutes the smaller, more vividly colored male fertilizes them with his sperm. The female thereupon scoops gravel over the pocket to keep the eggs secure. In a period of 19 to 80 days, depending on the temperature of the water, the eggs hatch.

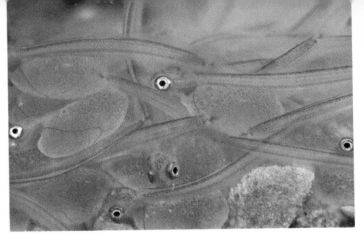

As newly born "sac fry" (above), young brook trout still have a yolk sac attached to their undersides, their only nourishment until they reach the age of 20 days, when the sac is absorbed and they are able to feed on microscopic animals. At the age of a year, the fingerlings are an average three inches long and mingle with adult trout (right).

Aquatic Lilliputians

Although the water of a pond may appear to be crystal clear, it actually teems with millions of tiny creatures, most of them invisible to the naked eye. These Lilliputians, the protozoans, rotifers, Hydras, medusas and water mites, are the quintessential pond creatures. Without them most animal life would disappear from ponds, and wetlands would become dismal, dead places. All of these minute creatures are important links in the food chain, all are invertebrates and all, with the exception of certain crustaceans, are completely aquatic animals that are born and spend their lives in the water.

The tiniest and most numerous of these pond dwellers are the one-celled protozoans, which, along with minuscule plants like algae, form the broad base of the pond's food pyramid. There are tens of thousands of species of protozoa, and they can be observed only by putting a drop of pond water under a microscope. Once revealed, their tiny world and their adaptations to it seem like something out of science fiction. Some protozoans anchor themselves to debris or vegetation by stalks. Others have whip or hair extensions that whirl in unison to propel them about and bring in food, which may consist of algae, decayed matter or other protozoans.

Among the most remarkable microscopic pond animals are the rotifers, or wheel animalcules, whose waving crowns of hairy cilia give them the appearance of little tops spinning madly around. Most are omnivores that cram food into transparent interiors, where it is chewed by grinding "jaws." Also visible within the rotifer's transparent body are its own developing young. Much larger than rotifers, though still generally less than an inch in size, are the Hydras. Freshwater cousins of sea anemones and jellyfish, Hydras are named after the many-headed monster of Greek mythology. They look like shreds of white or transparent string attached to rocks or water plants by one end, with the other frayed out into as many as a dozen filmy, waving strands. The strands are highly effective tentacles like those of an octopus that entangle almost invisible worms or water fleas on their surfaces and often immobilize the prey with poison "darts." Once they have entrapped a victim, the busy tentacles contract and move it into a single opening that functions as an entrance as well as an exit for any undigested remains. While a slight impulse picked up by the hydra's nerve network will cause the writhing tentacles to go to work instantly, a stronger stimulus may trigger a defensive reaction, causing the animal to shrink protectively into a small, jellylike blob.

Other tiny animals of the pond are more parasitic than predatory. Among them are various water mites, which, when they are young, are purely parasitic and become predators when they reach pinhead-sized adulthood. Under magnification mature mites look like bright-red beach balls equipped with two little eyes and four pairs of legs. While some swim or crawl busily about on their own to feed on even tinier animals, other immature mites attach themselves to such larger creatures as water scorpions or damselflies. Certain freshwater clams are also parasites, spending their larval stages attached tightly to the bodies of fish, which furnish them with sustenance until their growth is completed and they drop to the bottom to take up life on their own. The largest—and to any pond-swimming humans the most notorious—parasitic predators are the leeches, flattened, segmented worms that undulate through the sunless depths or attach their mouth and tail suckers to convenient surfaces to loop along, inchworm-style. Though some leeches are free-ranging scavengers or carnivores, many parasitize fish, snails and other hosts. Still other leeches, such as the large, dark-green, red-spotted *Macrobdella decora*, feed exclusively on the blood of vertebrates, including man, inflicting Y-shaped wounds with their sets of three sawlike teeth and gorging themselves before dropping off.

Some predators are included among about 30,000 species of freshwater crustaceans that inhabit ponds and lakes, although most are omnivorous scavengers of animal and plant debris. The smallest crustaceans are the water fleas and seed shrimps and such copepods as the one-eyed Cyclops. All are less than one tenth of an inch long. They feed on debris, algae and microscopic animals and are eaten in turn by fish. Somewhat larger, though seldom more than an inch in length, are the freshwater shrimp, many of which swim on their backs or sides. The giant of freshwater crustaceans is the crayfish, which is a familiar resident of ponds, streams, bayous and marshes of the Northern Hemisphere. Miniature versions of North Atlantic lobsters that rarely exceed five or six inches in length, crayfish generally hide by day and forage by night, crawling along on four pairs of walking legs and holding their long antennae and large pincers alert for random plant or animal food. If threatened, a crayfish shoots backward with surprising rapidity by contracting abdominal muscles attached to its scoop-shaped, lobsterlike tail.

Rotifer

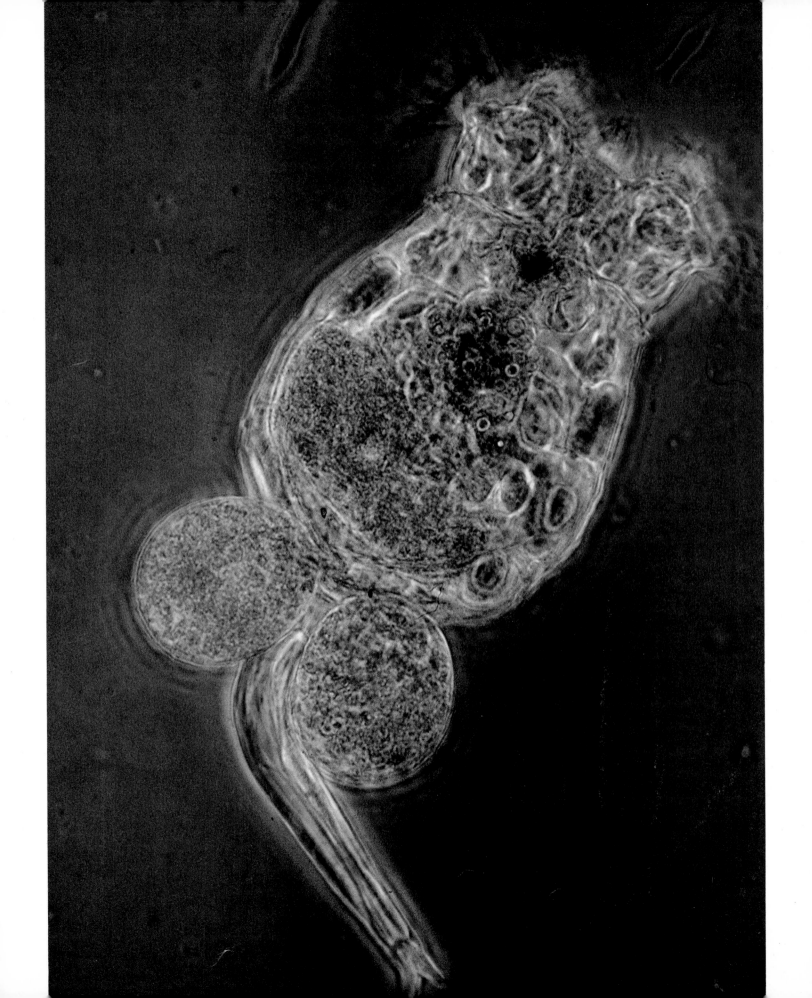

Freshwater Mollusks: Bivalves and Snails

The most familiar bivalves of the mollusk order—oysters, clams, mussels—have the briny tang of the ocean about them. But among the 10,000 species of bivalves, there are about 1,000 that inhabit ponds, streams, rivers and other bodies of fresh water. They have a remarkable degree of adaptability. Some inhabit only the purest, clearest water; others can tolerate ponds that are quite stagnant. Some tiny fingernail clams live in Arctic and Alpine pools frozen most of the year, but not many freshwater bivalves can survive in warm waters that exceed 90° F. One species, the European pearl mussel of Central European waters, was cultivated a hundred years ago for the pearls it produced but is now quite rare and of little commercial value.

The pond snails, cousins of the bivalves, are even hardier than the freshwater clams and mussels. Some breathe air through a primitive lung, enabling them to survive in stagnant waters—even in ponds that have been somewhat poisoned by industrial wastes—that other aquatic creatures could not endure. Because of the snails' special breathing apparatus, scientists theorize that they adapted to the pond by a circuitous evolutionary route as land animals that returned to the water.

The mollusk with the snaggletoothed smile (above) is a freshwater clam. Its "teeth" are actually tendrils arranged around the entrance of a siphon through which the clam filters in water containing oxygen and the tiny planktonic food that it consumes. The clam has another siphon through which it expels waste.

Most snails get oxygen directly from the water, but some species must return to the surface periodically to suck in oxygen through lunglike sacs. Pond snails are housed in two kinds of spiraled shells, one that is elevated and the other flattened in a "ramshorn" like that of the planorbid pond snail at right.

Like some transparent interplanetary space ship, Craspedacusta, a freshwater jellyfish—commonly called medusa for its fringe of poisonous food-gathering tentacles, which resemble the serpentine hair of the gorgon, Medusa, of Greek mythology—drifts in a pond (above). Jellyfish were believed to be exclusively creatures of the sea until a microscopic Craspedacusta was discovered in London's Kew Botanical Gardens in 1880—a stowaway from Brazil, it turned out, that had come into England in a shipment of tropical water lilies.

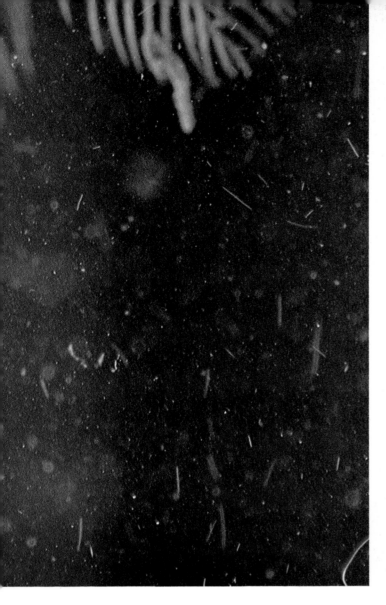

Invisible Masses

Cyclops, Hydra, Conochilus unicornus—the names have the ring of mythology about them, but there is nothing mythic about the minuscule animals, apart from their names, unless perhaps it is the fanciful forms they take. One of them—the starlike rotifer, *Conochilus* (below)—is actually a colony of tiny individuals that functions as a communal whole. Another, the *Craspedacusta* (left), is the only jellyfish with a freshwater habitat. *Cyclops* (left, below) is a copepod, one of a group of almost invisible crustaceans, and together with the rotifers they are the commonest of all pond animals, an essential food for fish and, with the exception of the protozoa, numerically the largest group of animals on earth, surpassing even the trillions of insects in numbers if not in species. Their importance in the food chain cannot be overestimated. And yet until the advent of microscopes no human eye could see them, and early scientists were not even aware that they existed.

Like the legendary Greek giant from which it takes its name, Cyclops (left) has just one magenta-colored eye. But the tiny crustacean is no giant, although its importance as a staple in the diet of many freshwater fish is enormous. Without the trillions of microscopic Cyclops in quiet streams and ponds, most freshwater fish, especially the schooling food fish, would disappear.

The curious formation at right is a rotifer colony composed of a number of wormlike animals attached to one another by elongated feet, with their eggs at the center of the "hub." The stronger worms are concentrated at the tip of each "spoke" and keep the colony upright by the beating of hairlike cilia on their heads that trap planktonic food and look like whirling wheels.

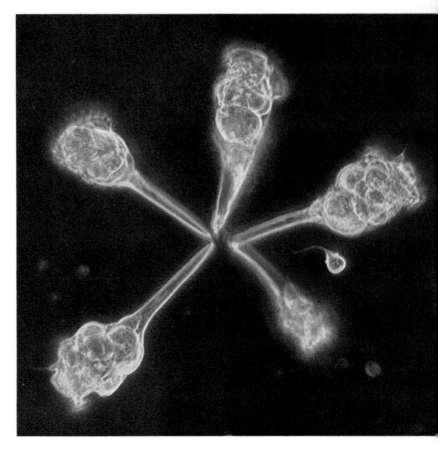

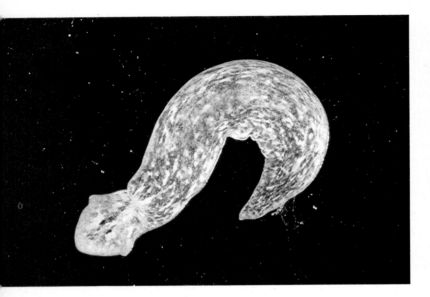

A flatworm (above) has a remarkable sensitivity to light and touch. In laboratories they have been trained to respond to shocks so that they can select the correct exit from a maze. They are regarded as menaces to fishing, devouring large numbers of fish roe and fry.

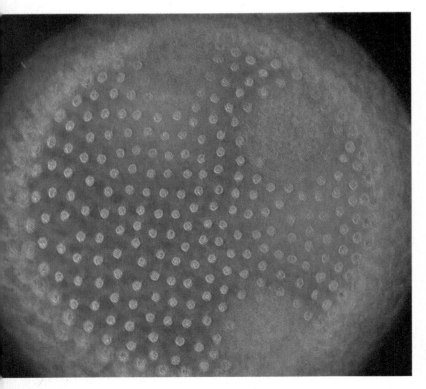

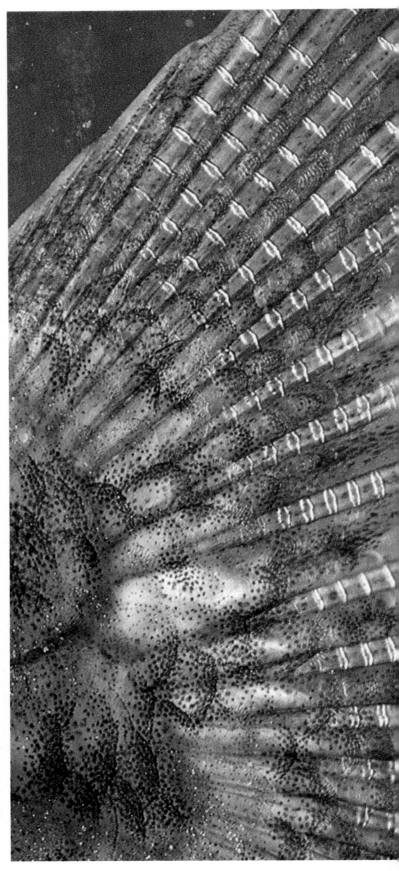

Looking rather like a diamond-studded blob of jelly, a Volvox (above) is a colony of 500 or more protozoans that moves by means of flagellates, long, whiplike hairs. Volvox are fish parasites, like the bloodsucking leeches at right that have attached themselves to the tail of a sunfish.

Scuds and Old Salts

The microscopic freshwater shrimp below is *Gammarus*, or "scud," a planktonic creature that abounds in unpolluted ponds, springs and streams of North America. Scuds (also called sideswimmers because they often roll over on their backs or sides) are voracious scavengers of almost any decaying matter on the bottom of their watery homes. Where population is especially dense—as many as 10,000 to three square feet of pond—they will attack living creatures or even cannibalize other scuds. They are in turn an important item in the diet of fish and other pond animals.

Their cousins, the brine shrimp (opposite), adapted to saltwater lakes, ponds and marshes millions of years ago, and their evolution is remarkable because their ancestors were freshwater creatures and not, as first presumed, from the sea. They are able to slough off any salts taken in with their food through the first 10 pairs of their legs, and, although their shells are impervious to the salt around them, brine shrimp of Europe are very sensitive to the degree of salinity in their environment. Depending on the salt content of their habitat, their appearance varies so radically that scientists were once convinced they were separate species.

With its curved back, flattened body, multiple legs and probing antennae, the translucent scud looks like its larger cousin, the oceanic shrimp. One subspecies, living in dark pools and streams of underground caves, is completely white, and its eyes have degenerated or disappeared altogether.

A pair of adult brine shrimp is shown at right, with the male hovering over a potential mate. The primitive half-inch crustaceans, which have adapted to life in the saline ponds and saltpans of the American West, are members of the order Anostraca, the fairy shrimps.

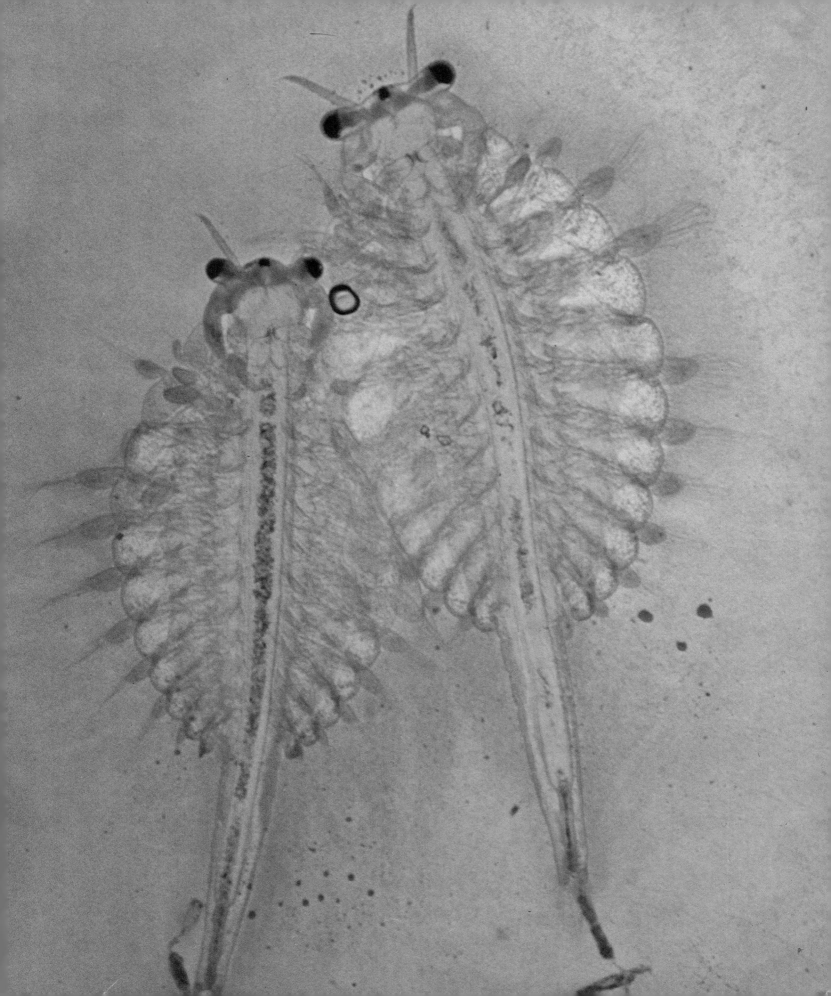

Resident Grouch

Crayfish, the largest of the freshwater crustaceans, are nevertheless relative runts—they grow no longer than eight inches—when compared with their oceangoing cousins, the lobsters. In their native habitat in the eastern and southern United States and in the Old World, where they are an introduced species, the succulent little creatures are a summertime table delicacy.

The only pond crustacean that ever ventures out of the water, the crayfish builds burrows with chimney-pipe openings (below) in the soft mud around the ponds and bayous where it lives and is able to scramble about awkwardly on the shore. Its most comfortable milieu, however, is in the water, where crayfish prey on smaller creatures and scavenge the pond bottom in search of carrion. Because they can administer a painful pinch with their claws, crayfish have a bad reputation among humans. Actually, crayfish will skitter away rapidly from any threat of danger.

With claws poised to pinch, antennae drawn back and tail tucked under its abdomen, a pale crayfish (right) assumes a defensive stance. When threatened, the shy creature can also propel itself backward to safety with a series of convulsive jerks of its tail.

Rising from the duckweed cover of a pond, the Louisiana crayfish at right looks like a special effect from a science-fiction film, waving its awesome pincers to threaten mankind. The photograph has been enlarged about four times. The crawdad could actually menace nothing more formidable than a tadpole or a water bug.

Amphibians and Reptiles

For most people the word "pond" conjures up a vision of a bullfrog on a lily pad or, perhaps, a pileup of painted turtles basking on a log. Indeed, the familiar resident amphibians and reptiles—frogs, turtles, tadpoles, snakes and salamanders—give almost any inland body of water its characteristic sights and sounds. The most audible of these are frogs and toads, with 70-odd species common to the United States and Canada. They provide the well-known voices of the pond in a symphony that makes the warm nights come alive with song.

Among the noisy singers is the shiny brown chorus frog. The male does all the singing, like all frogs, and is smaller than his silent mate—slightly less than one inch long, little larger than a human fingernail. Perhaps the best-known chorister is the spring peeper, known formally as *Hyla crucifer* ("cross-bearing tree frog") for the distinctive cruciform marking on its back. As winter tapers off with the first slight hint of warm weather the tiny males rouse themselves from hibernation, inflate their white throats into shiny, outsized bubbles and welcome the season with a high-pitched péeping so loud that it can be heard a mile away. The sound of several hundred peepers chorusing has been likened to sleigh bells or eerie pipes, but whatever it may recall there is no question that it is a herald of spring.

Just as common to ponds, and generally more visible, the three-inch-long green frog has a voice like the sound of a loose banjo string. Its song competes with that of the biggest of all American frogs, the bullfrog, which grows up to eight inches long and fills the summer evenings with a deep-throated *jug-o'-rum*. The bullfrog, like several other frog species, possesses huge external eardrums, which in the case of males are considerably larger than its eyes. Bullfrogs often drift splay-legged and motionless just below the surface, with only their snouts and bulbous eyes protruding, waiting for random prey—water beetles, dragonflies, smaller frogs, crayfish, even newly hatched ducklings and other small birds that come to the pond to bathe.

Quieter than frogs and more removed from the life of the pond, the toads spend most of their adulthood on land, although they begin their lives as gilled tadpoles and are inexorably drawn back to the water every spring to breed and lay their eggs. Salamanders are the other groups of amphibians most commonly encountered around ponds. According to species, they may live out their lives in the water or on the land or spend some time in both elements.

Among the reptiles that frequent ponds are the aquatic and semiaquatic snakes, such as water snakes and garter snakes that live along the shore and enter the water to feed on frogs, salamanders, tadpoles, crayfish, insects and worms. Most are harmless to man; the only venomous snake among them is the cottonmouth or water moccasin of the southern United States, which may reach five feet in length and usually opens its mouth wide to reveal a cottony white lining before it strikes.

The reptiles most typical of the pond are turtles, an ancient group that includes species ranging from the colorful painted turtles to spotted turtles, map turtles, mud turtles and musk turtles—the latter also called "stinkpots" for the strong odor they release when disturbed. Like most aquatic snakes, turtles are usually quite harmless to man, and many make satisfactory, undemanding pets. A dangerous exception is the common snapping turtle, whose sinister-sounding Latin name is *Chelydra serpentina*. It is among the largest and most thoroughly aquatic of the freshwater turtles, with a weight ranging from 10 to 25 pounds and occasionally 50 pounds or more. The snapper is easily recognized by its large head and spiny, alligator-like tail protruding from beneath a greenish-gray shell that seems at least one size too small. For its entire life *Chelydra serpentina* is a voracious eater, consuming aquatic plants, dead organisms and insects as well as fish, frogs, crustaceans and almost anything else that swims. It spends much of its time floating lazily just beneath the surface with only eyes and nostrils exposed. Usually it comes up beneath its prey, shoots out its long neck like a striking snake and grabs its victim with powerful, beaklike jaws. A young duckling, gosling or muskrat swimming on the surface is swiftly yanked out of sight and dragged to the bottom. Disturbed by an unwary or careless human on land, a snapper is quite capable of severing a finger or toe.

Somewhat less aggressive is the alligator snapping turtle, which inhabits lakes, rivers and bayous of the South. Most of the time it stays concealed on the bottom, its huge mouth gaping to reveal its "worm," a small red filament of flesh that it wiggles tantalizingly to lure hungry fish within reach. Largest of the freshwater turtles, alligator snappers average 100 pounds or more. The biggest on record tipped the scales at a ponderous 219 pounds.

Voices of Spring

More than birds, insects or any other creatures, frogs are the voices of the pond. Their vocal pipes ring out in songs of love when the males gather in and around the pond just after dusk—and sometimes all day long, especially at the height of the breeding season—to serenade the silent females and lure them into the nuptial pool. Tree frogs are the leading choristers, and their tintinnabulations are almost always heard in early spring or even, rarely, in the dead of winter.

After mating, the spring peepers and their brethren, the leopard and pickerel frogs, gradually fall silent, and the music is taken up through the summer and autumn months by larger frogs, the bullfrog and (below) the green frog, commonest of all Eastern and Southern pond frogs. The green frog's familiar baritone *gunk* differs from his love song and is, in fact, a territorial warning to other frogs.

A spring peeper (left) joins the male chorus, inflating his vocal sac like a wad of bubblegum. The cross that identifies spring peepers and gives them their scientific name, Hyla crucifer, *is visible on the tiny frog's back.*

His love song ended, a spring peeper returns to his customary arboreal perch (right), secured by the oversized suction toes that many tree frogs have evolved. Baby peepers are no larger than a fly, and when they reach maturity males measure just one and seven-eighths inches.

Besides singing along with the other romantic frog
choristers, the American green tree frog (above) has a
full repertoire of color changes, turning a buttery
yellow when singing and slate gray on cool days.
According to American folklore, it is an accurate
weather prophet, singing most frequently in the rain.

The gray tree frog (left) provides the reed section of
the hylid symphony, sounding its fluty call from a
choir loft in a tall tree. Its gray coat, sometimes
mottled with pale green, blends perfectly with the
lichens on the trees it inhabits.

Tuning up for a concert, a troupe of chorus frogs (above) sounds a rising, clicking love song. Tiny harbingers of spring, chorus frogs have vestigial finger pads, which mark them as onetime tree dwellers that have returned to life on the ground.

Pilgrim at Tinker Creek

by Annie Dillard

Annie Dillard, an amateur naturalist, devoted a year to the observation and study, with the occasional assistance of a microscope and books, of wildlife near her home in Roanoke, Virginia. Along with the beautiful and sometimes comical events and creatures she encountered she also came upon and faithfully recorded the grim side of nature in the raw. The vivid account of the grotesque death of a frog, the victim of a giant water bug, is excerpted from her book Pilgrim at Tinker Creek.

A couple of summers ago I was walking along the edge of the island to see what I could see in the water, and mainly to scare frogs. Frogs have an inelegant way of taking off from invisible positions on the bank just ahead of your feet, in dire panic, emitting a froggy "Yike!" and splashing into the water. Incredibly, this amused me, and, incredibly, it amuses me still. As I walked along the grassy edge of the island, I got better and better at seeing frogs both in and out of the water. I learned to recognize, slowing down, the difference in texture of the light reflected from mudbank, water, grass, or frog. Frogs were flying all around me. At the end of the island I noticed a small green frog. He was exactly half in and half out of the water, looking like a schematic diagram of an amphibian, and he didn't jump.

He didn't jump; I crept closer. At last I knelt on the island's winterkilled grass, lost, dumbstruck, staring at the frog in the creek just four feet away. He was a very small frog with wide, dull eyes. And just as I looked at him, he slowly crumpled and began to sag. The spirit vanished from his eyes as if snuffed. His skin emptied and drooped; his very skull seemed to collapse and settle like a kicked tent. He was shrinking before my eyes like a deflating football. I watched the taut, glistening skin on his shoulders ruck, and rumple, and fall. Soon, part of his skin, formless as a pricked balloon, lay in floating folds like bright scum on top of the water: it was a monstrous and terrifying thing. I gaped bewildered, appalled. An oval shadow hung in the water behind the drained frog; then the shadow glided away. The frog skin bag started to sink.

I had read about the giant water bug, but never seen one. "Giant water bug" is really the name of the creature, which is an enormous, heavy-bodied brown beetle. It eats insects, tadpoles, fish, and frogs. Its grasping forelegs are mighty and hooked inward. It seizes a victim with these legs, hugs it tight, and paralyzes it with enzymes injected during a vicious bite. That one bite is the only bite it ever takes. Through the puncture shoot the poisons that dissolve the victim's muscles and bones and organs—all but the skin—and through it the giant water bug sucks out the victim's body, reduced to a juice. This event is quite common in warm fresh water. The frog I saw was being sucked by a giant water bug. I had been kneeling on the island grass; when the unrecognizable flap of frog skin settled on the creek bottom, swaying, I stood up and brushed the knees of my pants. I couldn't catch my breath.

Salamander Cycles

The growth and metamorphosis of salamanders can take many forms. A few, like the Alpine salamander, are born live, but most originate as eggs in gelatinous masses. Some, such as the Mexican axolotl, spend their entire lives as gill-breathing larvae. Most, however, have a life cycle like frogs or toads. The process begins in early spring when adult female salamanders lay as many as several hundred eggs anchored to sticks or rocks in the water and swaddled in a gelatinous covering. These hatch as aquatic gill-breathing creatures that discard their gills by summer's end and leave the water to become lung-breathing terrestrial animals. A typical spotted salamander is shown below in a controlled situation with a clutch of eggs that is on the point of hatching.

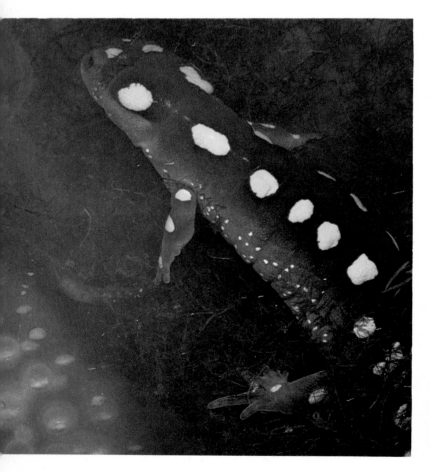

Like some ethereal moonchild, an embryo salamander floats in the globule of its egg (right). The gelatinous outer covering appears as a faint nimbus. Gills are already discernible, and before it hatches, tiny front legs will sprout, with the rear legs appearing later, after the hatchling emerges from the egg.

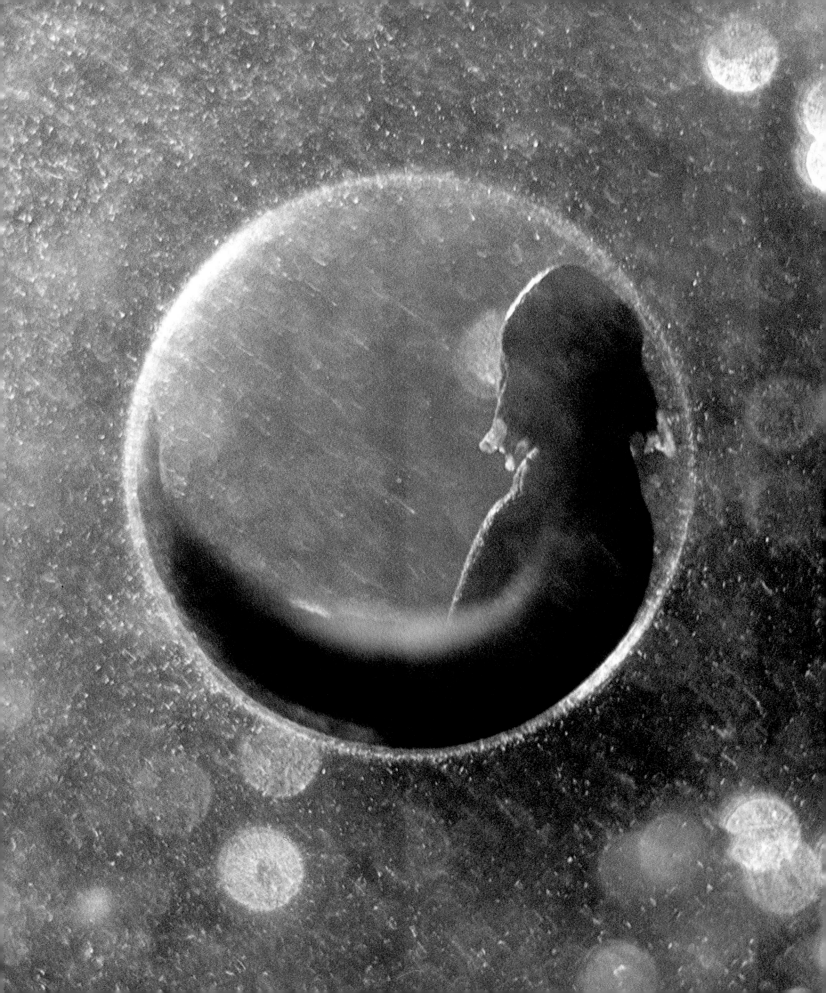

Water Serpents

Any snake will swim if the necessity arises, and many species are at home in the water. The water snakes, a large and cosmopolitan tribe, are truly aquatic and spend their lives in and around streams and ponds in every continent except South America. The diamondback water snake below is a native of wetlands of the middle-western United States south to Mexico. The diamondback is often mistaken for the cottonmouth water moccasin, the only poisonous aquatic snake in American waters, which shares much of the same watery habitat. Most water snakes, especially females, have hot tempers, however, and will deliver a painful, if nontoxic, bite when captured or cornered.

The closely related garter snake (opposite) is gentler and easily tamed. Its principal defense mechanism is an unpleasant-smelling liquid that it excretes when it is frightened. Garters are the most familiar, ubiquitous of American snakes, appearing in every one of the contiguous 48 states, Mexico and southern Canada in many variations of their black and yellow striped markings and many different environments. If they happen to live far from the water, garter snakes subsist mainly on worms, but if their habitat is beside a pond or marsh, they take readily to the water and become the local scourges of frogs and tadpoles. Although some species are quite at home in the pond, garters are generally less aquatic than their cousins. When threatened, a water snake will instinctively take to the water for refuge, but most garter snakes will head for the nearest grass or other cover on the shore.

Like a miniature Loch Ness monster, a common garter snake emerges from the duckweed to consume a fat tadpole. Although the tadpole appears to be too big to be swallowed whole, the garter snake can easily manage it with the aid of jaws that can be opened to a width of nearly 180°.

The diamondback water snake pictured above, basking on a log beside a Texas bayou, seems completely relaxed but is constantly alert for the hint of danger from birds or other predators and for any sign of a potential meal in the water—a frog, tadpole or small fish.

Small Pond Turtles

Some of the most colorful denizens of the pond, as well as a couple of the ugliest (pages 88–91), are the freshwater turtles. Most small American pond turtles are baskers with flattened carapaces and colorful markings, exemplified by the familiar painted turtle, with the characteristic splotches of scarlet on its shell, and the map turtle (right), with a row of dragonlike barbs down the center of its back. Cooters and sliders are characterized by yellow or red spots or stripes on their heads and necks and oversized webbed feet that are admirably adapted to swimming. The *Clemmys*—spotted turtles, wood turtles, Muhlenberg's turtles—have similar silhouettes and spend less time in the water than the other pond turtles. Mud and musk turtles have more rounded, drably colored shells and are the most aquatic of the small pond turtles, spending most of their time on the bottom, foraging for carrion, small crustaceans and fish. In times of drought, though, mud turtles will make long overland treks in search of new ponds.

The young map turtle above is easily identified by the barbed ridge down its back. As it matures, its shell will take on a pattern of contourlike lines, which accounts for its name.

The spotted turtles above belong to an extremely shy family. Although they are frequent residents of ponds, they also live in streams, marshes and other wetlands.

Cooters, like most pond turtles, are inveterate sunbathers (below), spending many hours absorbing the rays to increase metabolism and discourage the growth of algae on their shells.

According to Indian legend, the mud turtle (above) can chew its way out of an alligator's stomach. But its bite is actually no more than a nip, for it is the smallest (five inches) of all pond turtles.

Turtle Eggs for Agassiz

by Dallas Lore Sharp

When Louis Agassiz, the noted 19th-century zoologist and Harvard professor, was preparing his landmark work, Contributions to the Natural History of the United States, *he asked a certain Mr. Jenks, the principal of a small academy in Middleboro, Massachusetts, to find and bring him a clutch of mud-turtle eggs within three hours after they had been laid—and not a minute later. How Mr. Jenks undertook his unusual assignment is described in the following selection by Dallas Lore Sharp, a young naturalist. Happily, Mr. Jenks was able to accomplish his hairbreadth mission with only minutes to spare.*

"Then came the mid-June Sunday morning, with dawn breaking a little after three: a warm, wide-awake dawn, with the level mist lifted from the level surface of the pond a full hour higher than I had seen it any morning before.

"This was the day. I knew it. I have heard persons say that they can hear the grass grow; that they know by some extra sense when danger is nigh. That we have these extra senses I fully believe, and I believe they can be sharpened by cultivation. For a month I had been brooding over this pond, and now I knew. I felt a stirring of the pulse of things that the cold-hearted turtles could no more escape than could the clods and I.

"Leaving my horse unhitched, as if he, too, understood, I slipped eagerly into my covert for a look at the pond. As I did so, a large pickerel ploughed a furrow out through the spatter-docks, and in his wake rose the head of an enormous turtle. Swinging slowly around, the creature headed straight for the shore, and without a pause scrambled out on the sand.

"She was about the size of a big scoop-shovel; but that was not what excited me, so much as her manner, and the gait at which she moved; for there was method in it and fixed purpose. On she came, shuffling over the sand toward the higher open fields, with a hurried, determined see-saw that was taking her somewhere in particular, and that was bound to get her there on time.

"I held my breath. Had she been a dinosaurian making Mesozoic footprints, I could not have been more fearful.

For footprints in the Mesozoic mud, or in the sands of time, were as nothing to me when compared with fresh turtle eggs in the sands of this pond.

"But over the strip of sand, without a stop, she paddled, and up a narrow cow-path into the high grass along a fence. Then up the narrow cow-path, on all fours, just like another turtle, I paddled, and into the high wet grass along the fence.

"I kept well within sound of her, for she moved recklessly, leaving a trail of flattened grass a foot and a half wide. I wanted to stand up,—and I don't believe I could have turned her back with a rail,—but I was afraid if she saw me that she might return indefinitely to the pond; so on I went, flat to the ground, squeezing through the lower rails of the fence, as if the field beyond were a melon-patch. It was nothing of the kind, only a wild, uncomfortable pasture, full of dewberry vines, and very discouraging. They were excessively wet vines and briery. I pulled my coatsleeves as far over my fists as I could get them, and with the tin pail of sand swinging from between my teeth to avoid noise, I stumped fiercely but silently on after the turtle.

"She was laying her course, I thought, straight down the length of this dreadful pasture, when, not far from the fence, she suddenly hove to, warped herself short about, and came back, barely clearing me, at a clip that was thrilling. I warped about, too, and in her wake bore down across the corner of the pasture, across the powdery public road, and on to a fence along a field of young corn.

"I was somewhat wet by this time, but not so wet as I had been before wallowing through the deep dry dust of the road. Hurrying up behind a large tree by the fence, I peered down the cornrows and saw the turtle stop, and begin to paw about in the loose soft soil. She was going to lay!

"I held on to the tree and watched, as she tried this place, and that place, and the other place—the eternally feminine!—But *the* place, evidently, was hard to find. What could a female turtle do with a whole field of possible nests to choose from? Then at last she found it, and whirling about, she backed quickly at it, and, tail first, began to bury herself before my staring eyes.

"Those were not the supreme moments of my life; perhaps those moments came later that day; but those certainly were among the slowest, most dreadfully mixed of moments that I ever experienced. They were hours long. There she was, her shell just showing, like some old hulk in the sand along shore. And how long would she stay

there? and how should I know if she had laid an egg?

"I could still wait. And so I waited, when, over the freshly awakened fields, floated four mellow strokes from the distant town clock.

"Four o'clock! Why, there was no train until seven! No train for three hours! The eggs would spoil! Then with a rush it came over me that this was Sunday morning, and there was no regular seven o'clock train,—none till after nine.

"I think I should have fainted had not the turtle just then begun crawling off. I was weak and dizzy; but there, there in the sand, were the eggs! and Agassiz! and the great book! And I cleared the fence, and the forty miles that lay between me and Cambridge, at a single jump. He should have them, trains or no. Those eggs should go to Agassiz by seven o'clock, if I had to gallop every mile of the way. Forty miles! Any horse could cover it in three hours, if he had to; and upsetting the astonished turtle, I scooped out her round white eggs.

"On a bed of sand in the bottom of the pail I laid them, with what care my trembling fingers allowed; filled in between them with more sand; so with another layer to the rim; and covering all smoothly with more sand, I ran back for my horse.

"That horse knew, as well as I, that the turtles had laid, and that he was to get those eggs to Agassiz. He turned out of that field into the road on two wheels, a thing he had not done for twenty years, doubling me up before the dashboard, the pail of eggs miraculously lodged between my knees."

Curmudgeon

An acknowledged curmudgeon of the pond is, as its name indicates, the common snapping turtle. Although its cousin, the alligator snapping turtle (overleaf), is bigger, the common snapper is more aggressive. Even as leathery-shelled babies the size of silver dollars, young snappers will nip at almost anything that moves, although as hatchlings they are themselves quite vulnerable to the attacks of birds, snakes and other predators. Though they are among the most aquatic of turtles, patrolling the bottoms of ponds in search of food, snapping turtles do come out on land to lay their eggs and, in time of drought, to make lumbering migrations in search of another pond or stream. On such expeditions they seem even more irascible than they do in the water, lashing out at anything that crosses their path—as in the case of the elderly snapper shown below in the act of killing a water snake.

Snapping turtles will eat almost any living creature, but they are partial to carrion. The common snapper at right is dragging a dead fish to deeper water, where it will tear it to pieces with its powerful front claws and devour it. Like most freshwater turtles, snappers will eat only when submerged.

With a carapace resembling one of the great reptiles of the dinosaur age and a hooked beak like a bird of prey, the alligator snapping turtle (above) is the largest (up to 200 pounds) freshwater turtle and the biggest full-time resident of the pond. A native of southern wetlands of the United States, the alligator snapper is neither as aggressive—unless disturbed—nor as

active as its close relative, the common snapping turtle. In this photograph, the turtle is out of its element. An alligator snapper almost never ventures into shallow water, spending most of its life on the pond bottom, its mouth wide open, waiting for small fish to be lured by a wormlike red appendage on its tongue.

Aquatic Spiders and Insects

A pond on a late, warm afternoon often becomes a scene of feverish insect activity: iridescent dragonflies and damselflies hovering and darting on long, outstretched wings, gnats and midges dancing in clouds above the water, mosquitoes buzzing out on their nightly search for blood. Though often prodigious in their numbers, these highly visible pond denizens are only a fraction of the thousands of insect species—and their spider kin—that thrive in fresh waters, some living above the surface, others on it, still others in the murky depths. The insects furnish virtually all the vital sources of food for other pond animals—and even for a variety of carnivorous plant species.

Among the scores of insects that spend their lives floating on the water, using its tough, elastic film of surface tension to support their bodies, are the water striders, or "pond skaters," as they are sometimes called. Easily recognizable by a narrow body and slender legs—which depress the film into "dimples" that cast decorative shadows on a shallow bottom—a strider drifts motionless or skitters about rapidly, using its short forelegs to capture smaller insects and tiny animals that fall into the water or venture too close to the surface from below. Also a surface skimmer and hunter, the beautiful fisher spider weaves an ingenious "fishnet," not to trap insects, as other water spiders do, but to serve as a nursery for its young, which may number as many as 300.

Another creature that feeds on smaller insects that become trapped in the surface film is the whirligig beetle, which often congregates in groups that gyrate dizzily across the water, like so many Dodgem autos at a county fair. The whirligig is the only beetle that actually swims in the surface film rather than on top of it or below it. The upper parts of its body repel water, while the lower parts do not, allowing the remarkable beetle to float at midline. To enable it to see what is going on in both worlds simultaneously, its compound eyes are divided into two pairs; one pair looks up while the other looks down into the water. The whirligig is not confined to the surface; it can spread strong wings to fly if need be, and it can also dive deeply in search of prey, carrying a reserve supply of air like an aqualung in the form of a bubble trapped beneath its abdomen. The European diving bell spider also captures an air bubble in a silken sac, which enables it to stay submerged for long periods.

The water boatman is a skilled diver and strong flier, a slender inch-long creature that traps air under its wings and around its body in a silvery envelope, then propels itself downward by means of long paddle-shaped legs. Similarly equipped for scuba diving is the back swimmer, which spends its life upside down on its boat-shaped back, driving forward with large, oarlike hind legs and coming to the surface periodically to rest and stick its tail, equipped with water-repellent hairs, through the surface film in order to breathe. Such a snorkel technique is also used by predacious diving beetles, which hang head down with the tips of their abdomens projecting above the surface, and by the water scorpion, which lives underwater but replenishes its air supply by projecting a long tube from its tail up through the film.

Insects themselves are frequent prey of some of the strangest of the pond's predators, the plants that trap tiny animals for food. Growing in swampy shore areas where soil and water are acidic and low in nutrients, they supplement their nitrogen-deficient diets by capturing insects. The largest of the vegetable carnivores, often seen around the margins of boggy ponds, are pitcher plants, whose goblet-shaped, red-veined leaves secrete a sugary substance on their rims. Insects that are attracted to the nectar and venture inside are trapped on the leaves' flypaperlike sides. Eventually they drown in rainwater retained in the bottom of the plant's leafy cup, where enzymes reduce them to a digestible pulp. The smaller sundew plant attracts tiny flies and gnats to the sticky "rays" circling its half-dollar-sized leaves. Once trapped, the struggling victim is slowly enfolded by the leaves, which extract the vital juices from its body before opening again and allowing the remains to fall to the ground. The Venus's fly trap, a familiar, somewhat macabre house plant native to a small area of the Carolinas, functions in the same way, except that when trigger hairs on its clam-shaped leaves are touched it closes on its hapless victim.

The most aquatic of the meat-eating plants is the bladderwort, which resembles fine strands of lacy seaweed floating beneath the surface of the pond. The strands are studded with rows of minute, translucent bladders no more than an eighth of an inch in diameter, which in their normal position are empty, flattened sacs. When a water flea or other tiny organism touches a bladder's sensory hairs, however, its side walls spring instantly outward, briskly sucking the doomed insect in through an aperture with hairs. The victim swims frantically around inside until it finally succumbs and is absorbed.

Fisher spider

Eight-legged Aquanauts

The arachnid family includes a number of amphibious members that have made some remarkable adaptations to pond living. The fisher spider, *Dolomedes* (below), is an eight-eyed hunter that makes its home both above and below the surface of the water. The dense hairs covering its body serve a two-fold purpose. They prevent the fisher spider from breaking the water's surface film by helping to distribute its weight evenly over a large area. And, because they are water-repellent, the hairs trap air bubbles when the spider submerges, providing enough oxygen to allow the spider to breathe underwater for an hour.

The diving spider (opposite, below) spends almost its entire life in air-filled bathyspheres it constructs beneath the surface. A fine web is anchored to aquatic plants. Then, one by one, several air bubbles are trapped on the surface, brought down and released under the silken net, making it balloon upward like a dome. It is in such structures that the spider eats, sleeps and even reproduces.

The red water mite (opposite, above) is a pinhead-sized relative of the spider. It can be found skittering through the water or making its way along the bottom transported by four pairs of hair-fringed legs.

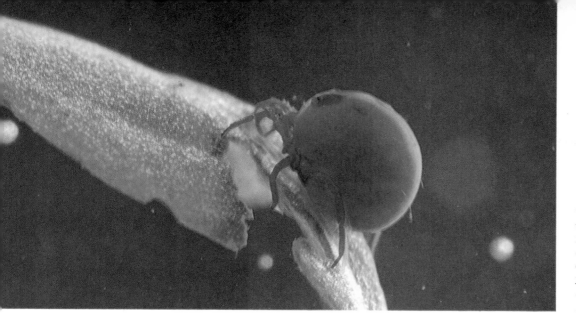

This crimson balloon-shaped body set with two coal-black eyes belongs to the red water mite (left). The tiny creature is a year-round pond dweller, braving the coldest weather of winter. Adult mites feed mainly on plant matter, while their larvae parasitize numerous other pond-dwelling animals.

Its hairy legs splayed out around it, a fisher spider (left) rests quietly on the surface of the water, waiting for an unsuspecting insect to come along. Fisher spiders rely completely on their hunting skills to catch their food, spinning a web only when the time comes to hatch and rear their young, which may number 300 spiderlets.

A diving spider, Argyroneta aquatica, lives head down in its underwater home. The web that contains the air bubble is not used to entangle prey. But if an enemy or suitable prey happens to stumble across one of the lines, its vibrations are transmitted to the spider, alerting it to the situation.

Walking on Water

The water strider is a familiar sight at a forest pond or the edge of a sluggish stream, skating across the water toward some insect prey that has fallen to the surface or trolling about slowly in groups, dappling the bottom with clover-leaf shadows (opposite) that are reflections of the dimples its feet make on the water's surface film.

The secret of the strider's ability to walk on water is in its long legs, which give it a delicate balance on the film, and in its specially adapted feet, which end not in claws, like those of most other insects, but in tiny tufts of hairs that support it on the membranelike surface tension of the water. Water striders can even leap into the air and land again without penetrating the surface tension. Should a strider accidentally become submerged in still water, minute velvety body hairs, coated with a waterproofing oil, trap air bubbles that instantly refloat it. In rapidly moving waters, though, a strider may drown.

Above, water striders gather over the body of a fallen dragonfly. Sensory hairs on the strider's legs detect the struggles of a downed insect at considerable distances. Once in the grip of the strider's shortened forelegs, the victim is sucked dry of its vital juices.

Legs extended like the outriggers of a boat, a water strider (top) dimples the water but does not break through. Though they have wings, striders are feeble fliers, but they can leap forward on the water almost a yard with a single thrust of their legs.

NEAR HORIZONS

BY EDWIN WAY TEALE

The world of insects is no farther away than the nearest backyard but is overlooked by most people. Edwin Way Teale, a distinguished naturalist, writer, photographer and past president of the New York Entomological Society, has explored that world and described it meticulously in Near Horizons, *a volume he calls "the travel book by the man who stayed home." During his explorations he once shared his lunch with a group of water striders and thereupon wrote an account, excerpted here, of their busy activity on the surface of a pond.*

But the main interest there was not in the visitor but in the regular inhabitants, in those swift, slender-bodied, spider-legged insects which use the surface film for their hunting ground.

Most of the time these water-striders drift about with only an occasional flip of their long oar-like middle legs. These legs form their driving mechanism. Their hind legs steer like rudders and their forelegs, carried raised and waiting, capture their prey. Fine hairs form a velvety coating over all the limbs. This pile prevents them from becoming wet. Once wet, the legs would plunge through the surface film.

While these dwellers on my cove drift about in seeming indolence for hours at a time, they can come to life with the suddenness of an explosion. In sight of prey, they leap and pounce like a cat on water. I have seen one of them jump fully an inch and a half into the air. It landed on the surface film again and went darting off, a quartette of golden spots

racing behind it over the black bottom of the shallow stream. These spots were formed by rays of sunshine refracted in passing through the curved dimples at the insect's feet.

As water-striders drift on the surface film, they often spend long periods cleaning themselves. Once one drifted close to my peninsula with only three dimple-spots showing on the mud bottom. It was standing on three legs while it carefully cleaned one of its middle feet.

On days when I come to the garden with a sandwich stuffed in my pocket, I always share the meat with the water-striders. A grand mêlée follows the splash of a fragment of ham or bologna tossed into the cove. The water-striders ignore bits of bread. But no matter what the meat, they attack it like wolves. Once fully fifty striders clustered in a solid mass about a lengthy strip of spiced ham. In the excitement, a newcomer leaped completely over the mass. Another time, one strider straddled a slender bit of ham and, like a tugboat laboring in front of a barge, dragged the meat—with nearly a dozen other striders clinging to it—half-way across the cove.

At sunset, on another day, a water-strider leaped on a piece of meat almost as soon as it struck the water. A score of other striders bore down upon it from all directions. Grasping the meat in its forelegs, the first insect darted ahead. Like a halfback carrying the ball through a broken field, it dodged and twisted, whirled and sidestepped. Twice it almost lost its prize. But at the end of a strenuous minute, it had shaken off its pursuers and was pulling the fragment of meat up on the mud of the bank. Here it was later joined by two others. For more than an hour, the three insects fed, hardly moving and overlooked by others of their kind.

Neither ham nor bologna are on the normal menu of my water-striders. Their main source of nourishment comes from the vital juices of small insects which float on the surface of the cove, many of them dropping from the overhanging flags and willow.

Such windfalls descend almost continually on windy days in summer. I remember once when a shower of minute tree-hoppers and other Lilliputian insects pockmarked the whole surface of the cove each time a gust shook the branches of the willow. The water-striders darted this way and that, seizing a struggling victim, draining away the fluids from its body, sweeping on in search of other prey.

Most of the food of the water-strider is comprised of insects smaller than itself. But occasionally a moth or a fly is trapped on the surface of the water to provide a special feast. I remember seeing several water-striders, coming from different points of the compass, fall upon the body of a large crane fly which had ceased its struggles and lay quiet on the surface of a rock-enclosed pool in a woodland brook in Maine. At this same miniature lake, hardly more than a foot and a half across, I saw, on another day, twenty-eight water-striders all facing in the same direction like a flotilla of slender cruisers. With regular, deliberate kicks of their middle legs, they maintained their position in the flowing water just beneath an eight-inch waterfall. All were waiting for the current to bring them food.

KSENIAK

Beetle Submariners

Prized by collectors for their beauty, vivid color and often unusual form, beetles are the most familiar creatures in the insect world. It is estimated that one out of every four known species of insect is a beetle—a vast host of approximately 280,000 species. Of this number relatively few live in water. Common to most lakes and ponds are the predaceous diving beetles of the family Dytiscidae and the gyrinid beetles, popularly known as whirligigs.

One of the largest and most active of insect predators, the diving beetle grows to a length of over an inch and feeds on tadpoles, small fish and other insects. The larva, commonly called a water tiger, is also a predator and captures prey with the help of its hollow, razor-sharp jaws. In the picture at right of a water tiger devouring a tadpole, the jaws are pumping lethal digestive enzymes into the victim that not only kill it but also liquefy its body contents, which the larva thereupon sucks up through its hollow jaws. There are over 300 species of diving beetles in North America. Whirligig beetles get their name from their dizzying method of swimming, usually in crowds in a maze of circles, at such breakneck speed that it is nearly impossible to track an individual's course. They make their home on the surface film of the pond—the juncture of air and water—but they are also quite capable divers. Only the upper surface of the whirligig's plump body is water-repellent and buoyant, so it swims half above and half below the surface. Adult whirligigs are scavengers, but their larvae are voracious carnivores.

As a whirligig beetle skims across the surface of a pond (left), its wide, flat "rowing" legs give it an extraordinary swimming ability. But what really makes the whirligig a curiosity of the insect kingdom is its remarkable eyesight. Its eyes are divided into two separate pairs—one above and one below the water—which enable the skittery beetle to spot its victims and to avoid its foes at the same time.

The giant diving beetle (below) is able to breathe and maintain its specific gravity under water by means of a supply of air it carries between its back and wings. With such specialized equipment, the predatory beetle can rise or sink at will in its everlasting hunt for living prey, which it kills with daggerlike mandibles.

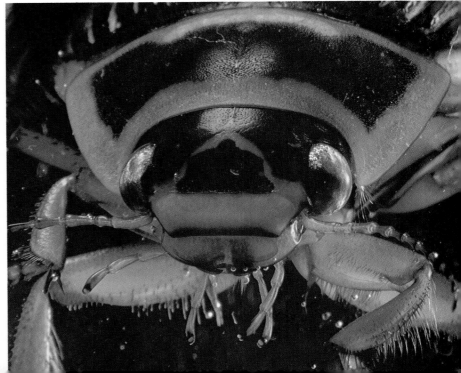

Brief Encounters

The familiar fishbait known as "crawlers" are actually the larvae of dobsonflies, denizens of fast-flowing freshwater streams. Immature dobsonflies spend two to three years creeping and swimming about the stream's bottom in preparation for an anticlimactic adulthood. Mature males live no more than one to two days, females from one to two weeks, or just long enough to lay eggs.

The life expectancy of caddisflies (above), another important food source for freshwater fish, is no greater. Like dobsonflies, they have an abbreviated adult stage lasting from a few days to several weeks, after several years as aquatic larvae. Caddisfly larvae build unique protective cases for themselves from a sticky silk that they produce. Then they disguise the tiny cases with sticks, leaves, sand, pebbles or whatever else may be at hand. Caddisflies develop in fresh water almost everywhere in the world.

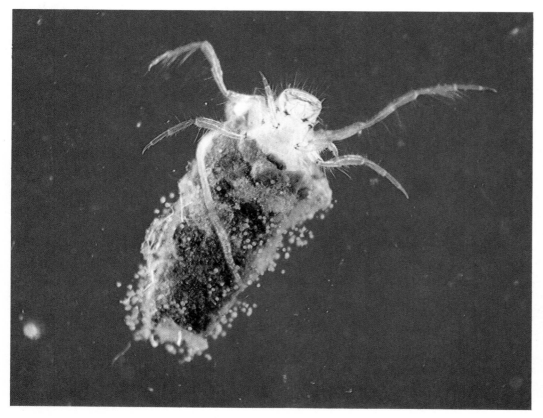

Most of a caddisfly's life is spent as a larva (left). As winged adults (above) they live only long enough to procreate—a matter of hours—before dying. The grotesque creature at right is a female dobsonfly. Males have even larger, more formidable-looking pincers, which serve one purpose alone: to clasp the female during their brief mating. A female (top, right) prepares to lay her eggs—the last act of her life.

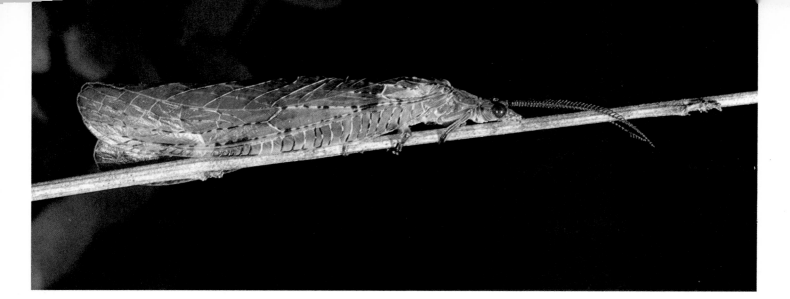

Dragon and Damsel

Wherever there are inland waters there are likely to be dragonflies and damselflies. These most delicate insects have been on earth since before the dinosaurs; fossil remains of ancestral dragonflies date back over 300 million years. Sole members of the order Odonata, dragonflies and damselflies are large predatory insects. The major portion of their approximately two-year life-span is spent in the water as larvae that hungrily pursue water insects. After molting two or three times, allowing for the lengthening of their bodies and the development of their two sets of wings, dragonflies and damselflies take to the air. Aided by superb eyes that enable them to see as far away as 40 yards and wings that carry them as fast as 30 miles an hour, they are as successful at hunting in the air as they were under water.

A dragonfly larva, also called a nymph, gobbles up a worm (above), just part of its daily intake. Dragonfly larvae are aggressive, voracious predators that scour the bottoms of ponds, lakes and riverbeds in search of food. They are equipped with extensible lower lips, called labia, that are adapted for catching prey.

Its diaphanous wings outstretched from side to side, an adult dragonfly (left) pauses on a reed before resuming its quest for likely prey. Adult dragonflies spot prey visually, catching insects on the wing and crushing them with their powerful jaws. Dragonflies' eyes are the largest in the insect world, each orb possessing as many as 28,000 facets.

A delicate damselfly alights on a stalk of sundew to look for insect prey (opposite). Unlike its larger cousin the dragonfly, the damselfly does not eat in the air but prefers to rest on vegetation as it feeds. Damselflies are also able to keep their wings folded behind them when perching, a feat a dragonfly cannot execute.

Circes of the Bog

In an era of horror films some sharp movie mogul will undoubtedly produce an acetate adaptation of the myth of the "man-eating tree of Madagascar," although it is certain that there are no carnivorous plants capable of devouring a human being. There are, however, some 450 species of carnivorous plants—the Circes of the bogs—that have developed ingenious traps in which to ensnare unsuspecting insects. The most elaborate is that of the bladderwort. This aquatic plant bears on its branches tiny bladders that are kept closed by a valve surrounded with sensitive hairs. When an unwary insect touches one of these hairs, the valve opens wide and the bladder expands, vacuuming up water and victim as well. Then the valve closes and the doomed insect is digested.

The exotically beautiful, trumpet-shaped pitcher plants, so innocently reflected in a pond in the New Jersey Pine Barrens at left, do their prey in with the utmost efficiency. Their hollow leaves contain a sweet-smelling liquid consisting of rainwater and a fluid excreted by the plant. Attracted by the scent of the nectar, insects come to taste the brew and are trapped in a lining of sticky, downward-pointing hairs. Enzymes and bacteria in the water slowly decompose and digest the hapless insects which in turn nourish the bloodthirsty plant.

The crimson pitcher plant above is a death trap for most insects (right). Some species of mosquitoes, flesh flies, caterpillars and a species of ant boldly make their homes in the lips of the plants and nowhere else. It is believed that such insects are immune to the digestive action of the enzymes in the plant's fluid.

A mosquito larva appears to be nesting comfortably in the cradle of a bladderwort in the controlled situation below. But if it were smaller it would be doomed. The bladder can suck a victim in with such incredible speed that neither the human eye nor a camera could record the action. And 20 minutes later the bladder will be reset and ready for its next victim.

The rash fly at right is about to walk into the parlor of a killer far more treacherous than a spider. It is inching its way along the leaf blades of the Venus's flytrap. The lethal plant's leaf is divided into two halves, which are hinged in the middle like a change purse. The leaf will not close unless two of three sensitive hairs that line each half are touched in succession or the same hair is touched twice. But once it closes, the hapless fly will be finished.

Birds of Pond and Marsh

Ponds are the natural habitat of many kinds of birds, which seek out the water for a drink, a bath, a source of abundant food and a nesting site on the shores or in the thickets and trees nearby. Usually the most evident are summer residents, the water birds—myriad ducks, geese, grebes and occasional loons and swans that splash into ponds and lakes during their annual migrations, as well as herons and other long-legged waders that stand like sentinels or patrol the shallows looking for a meal of frogs, fish or small crustaceans. These are only the most spectacular of the pond birds; the swarming life in the water, and particularly the clouds of insects and their waterborne larvae, attract vast numbers of other birds as well.

Among the smaller birds are certain species of swallows, wrens and warblers, including the water thrush and the brightly colored prothonotary warblers. One of the most conspicuous of pond birds is the red-winged blackbird, which, in various subspecies, is a noisy denizen of North American wetlands from coast to coast. Along with spring peepers and chorus frogs, the redwing is a familiar herald of spring. In northern areas males often arrive from southern wintering quarters to scout the likely nesting areas while the snow is still on the ground. Flashing their distinctive gold-tipped red epaulets, they raucously proclaim their territories and defend them against other birds. The females, dressed in drab, brown-streaked plumage, join their mates later in the spring to build nests among the thickets of reeds or cattails and shrubbery near the open water, where they lay clutches of three to five eggs and catch great quantities of insects to feed their young.

Not always as visible but equally characteristic of the pond's swampy margins are several species of sandpipers, snipes and rails. Among the shiest and most retiring are the reed-thin Virginia and sora rails, which conceal their nests in the tall marsh grasses. When flushed out by an intruder they run swiftly through the underbrush or spring into the air and fly feebly for a few yards with long legs dangling. Then they drop, seeming to feign exhaustion, to lose themselves in the reeds again.

Equally shy and slender, bitterns employ an added protective technique when surprised in their marshy habitats: They freeze in position with their long bodies extended and their pointed heads and bills sticking straight up in the air; the stripes on their necks and bodies blend deceptively with the vertical patterns of light and shadows among the reeds. If a breeze should set the stalks rustling while a bittern is holding its camouflage posture, the bird may sway slowly from side to side with its moving background, stopping as the wind subsides and standing as rigidly as before. If a bittern is threatened while it is nesting, though, it may reverse its tactics, puffing itself up and flapping its extended wings in an attempt to frighten the invader away.

Other birds that feed on the animal life of ponds include various hawks, notably the magnificent fish-eating osprey, and kites, among them the graceful swallow-tailed kite and the rare Everglades kite. A native of South America and southern Florida and a fastidious feeder, the Everglades kite subsists almost exclusively on one species of freshwater snail, which it pries from the shell with its long, hooked bill. Probably the most widespread—and in its own way the most spectacular—of the flying fishermen is the kingfisher, whose various species frequent ponds, lakes and streams over much of the world. The only one in North America is the belted kingfisher, which is only slightly larger than a robin, with a disproportionately large head, surmounted by a ragged, gray-blue crest that rises above a short tail and tiny legs. Despite its top-heavy appearance, the kingfisher is a masterpiece of functional design. Its big head and beak form a sturdy, spearheaded instrument that is built to withstand the shock of frequent dives and jarring impacts in pursuit of fish. Perched on a prominent tree branch where its view is unobstructed—or hovering like an oversized hummingbird 20 to 40 feet in the air on rapidly beating wings—it scans the water for a target, then abruptly plunges downward in a straight or spiraling dive, momentarily disappearing beneath the surface with a splash. After catching a fish in its bill, the kingfisher flies back to a favorite perch, beats its prey senseless and deftly maneuvers it into a head-first position so it can be swallowed without injury from the spiny, rear-pointing fins. The kingfisher seems to relish its role as avian king of the pond, flying swiftly and confidently as it patrols its territory, swooping up to alight on a branch and utter its cocky battle cry—a loud, harsh, rattling call that resembles nothing so much as an old-fashioned wooden noisemaker used to celebrate New Year's Eve.

Red-winged blackbird

Ubiquitous Redwing

The sedges, reeds, cattails and shrubbery that border ponds, streams and marshes from Central America north to the Yukon and Nova Scotia are nesting grounds of the red-winged blackbirds. Each spring some redwings fly north to mate and nest, while others remain in the south and abandon their homes in fields to migrate only as far as the nearest wetlands to establish their nesting colony.

The females alone incubate the eggs, which hatch in 11 to 12 days. In July, when the young are ready to travel, the redwings regroup into flocks that feed, often destructively, on ripening grain. Crop losses are estimated at between $20 and $30 million annually. Redwings make up for at least part of the havoc by consuming large quantities of insects, such as beetles and caterpillars, that are even greater crop destroyers.

The flashing red and yellow epaulets of the adult male red-winged blackbird (above) make him easily identifiable. The gender of hatchlings (below) is not so easy to identify. All young are brown, as are adult females. Young males develop mottled shoulder patches that eventually turn scarlet.

A male redwing (above) swoops down menacingly toward a little blue heron that has perched too close to its nest. The male remains protectively nearby, defending his territory, his mate and their brood, although the female red-winged blackbird takes on all other responsibilities for her young. She builds the nest, incubates the eggs and feeds the hatchlings. After the breeding season is over the redwings desert their nesting colonies, flying off in segregated flocks—males in one (left), females and young in another—that may number hundreds of thousands of birds.

113

Hardy Migrants

Among the most familiar seasonal sights around the pond is that of approaching flocks of noisy ducks and geese making a feeding stopover on their way to winter havens. The most widely distributed and perhaps best known of these migratory waterfowl are the mallard ducks and the Canada geese. Mallards are hardy creatures capable of enduring the coldest winter weather as long as open bodies of water and food are available to them. While the greatest concentrations of wintering mallards are found along the Gulf Coast and in the Mississippi Valley, flocks of the handsome birds remain as far north as Alaska throughout the year.

Similarly, some Canada geese stay in the North during the winter. But for others, the season's shortening days and increasing chill signal the beginning of a journey that may take them as many as 3,500 miles south to winter refuges in Mexico. There they stay until early spring, when they reverse their routes and follow an unerring homing instinct back to the same breeding grounds in the North where they themselves were hatched.

Like guests at a skating party, a flock of mallard ducks strolls across a frozen pond in Pennsylvania. The splendid green and chestnut winter colors of the males contrast dramatically with the females' brown plumage. In the picture above, a spray of water settles around a boldly patterned Canada goose as it makes a splash landing in a New Jersey pond.

Nesters at the Pond

The profusion of reeds and tall grasses bordering ponds and crisscrossing marshes provides excellent camouflage and protection for the many species of birds that nest in the wetlands. One of the regular nesters is the common gallinule (right), which securely anchors its shallow nest of dry cattail leaves and rushes among the marsh vegetation, several inches above the water. Even more commonplace around North American ponds is the ducklike coot, which is similar to its common Old World cousin except for white hash marks on its sleek black wings.

The brightly plumed Louisiana heron (below, left), one of the most numerous herons of the southern United States, breeds in colonies. Both male and female incubate the three or four blue-green eggs that are laid on a stick nest placed either in a tree, in the grass or among reeds. Two other herons, the least bittern (below, right) and the American bittern (opposite), lay their four or five eggs on dried cattail or bulrush stalks that are bent to form nesting platforms. Once hatched, the young are fed on semidigested fish and crustaceans regurgitated by their parents.

Its yellow-tipped red bill shield contrasting sharply with its sleek slate-colored body, a common gallinule floats serenely on the water's surface. The common gallinule, also called the common moorhen, is found on every continent except Australia and Antarctica.

A Louisiana heron perches on a thin branch overhanging the water, looking for the aquatic creatures it feeds on. The graceful long-necked, long-legged wading birds frequent shallow pond waters in search of frogs, fish and crustaceans.

A least bittern (above) tiptoes daintily across a lily pond. Like its larger cousin, the American bittern (opposite), this small heron is a shy, reclusive creature. When disturbed, it freezes to blend in with the reeds or flees on foot rather than taking to the air.

Moose and Deer

Many animals live outside the ecosystems of ponds but must visit them regularly for the water that sustains all life. Among these visitors, perhaps the shiest and loveliest are the deer; the most imposing are moose. More numerous are smaller mammals—field mice, rats, rabbits, an occasional porcupine—that venture furtively to the shore to drink their fill and then scurry back to the shelter of their accustomed habitats. Others linger on to look for food as well as water at the pond. Skunks in particular spend hours searching for the eggs of turtles and wetland birds before returning to higher ground. Foxes, coyotes and bobcats arrive to slake their thirst and to pounce on any unwary rodents or other animals they may flush. Bears may splash through the shallows hunting for fish or beavers.

Early morning and late afternoon are the chosen times for the visits of deer, which usually wander down to the water singly or in herds. They drink deeply, then feed on marsh marigolds, young ferns and other tender plants that grow around the pond, wading out up to their bellies to browse around, searching for aquatic shoots. Often the deer splash into deeper water for a cooling dip or to find relief from biting insects. In an emergency the water may be a haven from pursuers, a place where the long-legged deer can even the odds and make a stand or swim swiftly to safety on an island or far shore. Deer visit ponds and marshes at all seasons, including winter, when herds unable to find enough browse in frozen uplands come down to the valleys to strip off the shrubby undergrowth. Their narrow, dainty hooves are sometimes a winter hazard, however, and deer have great trouble maneuvering on ice. If pursued across a pond's frozen surface, they may easily slip and fall and are often helpless to regain their footing.

A rare sight at most ponds, though not uncommon in northern wilderness areas through North America, Europe and Asia, is the giant of the deer family, the moose. The big, gangly animals single out favorite ponds in their territories and visit them daily, usually in early morning and evening, either singly or in family groups. In the summer, when the black flies and mosquitoes are biting, moose may stand in the water up to their necks for long periods to escape the torment, or roll in a pondside wallow to cover themselves with a protective coating of mud. In hot weather the gangling young calves and even adults sometimes splash around playfully in the water, chasing each other like oversized children at a local swimming hole.

Moose do much of their feeding at ponds, where dense stands of succulent young willows, aspen and birch crowd the shoreline. They also crop the grasses and sedges of pond meadows, as well as ferns, mushrooms, wild parsnips and other plants. A moose may spend as much as an hour at a time foraging in the water for the tender leaves, roots and tubers of such aquatic plants as pondweed, wild celery and wild rice, sometimes submerging its head for as long as a minute, then lifting it with a snorting noise and a splash of cascading water, mouth full of dripping vegetation. Moose are remarkably adept at finding their favorite foods underwater and will sometimes dive completely below the surface in deep water to get at them.

Spending as much time as they do in and around freshwater, moose have become powerful, natural swimmers. Calves are urged to paddle by their mothers when they are only a week or two old. A large adult moose can swim as fast as six miles per hour, a speed equal to the best that beavers and otters can manage, and keep it up for two hours at a stretch, outstripping even an experienced backwoodsman in a light canoe. Unlike deer, moose have little fear of ice, on which their large cloven hooves find a secure foothold. As soon as the northern ponds and lakes freeze solidly, moose move confidently across the ice. In summer their long, slim legs can carry them surefootedly over tangles of fallen trees, and their hoof pads can spread wide to give them support in soft or muddy ground.

However well it has adapted to its habitat, the moose is a homely, unlikely-looking animal. Its stiltlike legs seem too thin for its massive body; its large forequarters and slim hindquarters appear to be designed for two different creatures. Its tail, less than three inches long, looks like an afterthought of nature and is of no help at all in swishing off flies. It has big ears, small eyes, the pendulous nose of a slapstick comedian and an oddly useless dewlap of skin called a "bell" dangling from its throat. Yet for all its homeliness a moose is an undeniably dignified, even majestic, animal. A big bull with a massive, scooplike crown silhouetted alertly against the sky is the largest antlered animal in the world, towering over all other wetland creatures as the ultimate lord of the pond.

Bull moose

Wetlands Monarch

With their huge dish-shaped antlers rearing a full ten feet above the water, moose are among the most impressive of all the animals that frequent ponds. Multitudes of them once roamed the Northern Hemisphere, but the depredations of man thinned their ranks to the edge of extinction. With careful conservation programs, however, these largest of all antlered animals have made an impressive comeback. In 1963 there were an estimated 120,000 moose in Alaska—double their 1960 population.

In late May or early June, cows give birth to one or two calves. After about a week the young are strong enough to tag along behind their mothers. The females and their calves stay near the water much of the summer. The bulls seek higher altitudes to escape the torment of biting insects, returning to the pond and the females at the next rutting season. The calves remain with their mothers until the next spring, when the cows, in preparation for the arrival of new calves, drive the yearlings away, forcing them to fend for themselves.

The Alaska moose cow in this sequence of pictures demonstrates (from left) her semiaquatic method of feeding on bottom-growing pond plants—breathing, ducking for food and emerging for another breath. Moose have been observed diving into water more than 18 feet deep to bring up food.

120

A moose cow shepherding her two young calves across a stream is not an uncommon sight in Alaska's McKinley Park in the month of June. Unlike most other young deer, which are generally spotted at birth, moose calves are a uniform reddish-brown color. As a calf gets older its coat becomes blackish brown to brown in color, like its mother's.

Whitetail's Haven

During the late spring and early summer, the wetlands of the Western Hemisphere, from southern Canada into South America, are favorite haunts of the white-tailed deer. There the graceful, shy animals find a cool refuge from the mosquitoes, ticks and deer flies that plague them, as well as an abundance of aquatic vegetation such as pond grasses to feed on.

The white-tailed deer is named for its long white tail, which it erects at the first sign of danger and waves back and forth like a warning flag to other deer as it flees (overleaf). The buck's antlers, which consist of a main horn with smaller points forking out, are other distinguishing features. Antlers, shed every winter, begin to regenerate by April or May. By the time mating takes place in November and December, they are fully regrown and are used as weapons in the noisy battles that occur daily between rival bucks over mature does. Through no coincidence, the annual hunting season occurs at the same time, and their magnificent antlers make white-tailed bucks prime targets for shooters.

A group of white-tailed deer—males, females and young—fords a bayou in Louisiana. Although these particular deer do not generally congregate in herds, it is not uncommon to see small family groups traveling together. Numbering from two to four animals, these bands are usually made up of a doe and her young.

On spindly legs, a male white-tailed fawn stands partially obscured by the foliage bordering a lake in Colorado. Lacking the stamina to keep pace with their mothers until they are a few weeks old, fawns are usually left in the protective cover of dense undergrowth while their mothers feed, lying low until the parent returns to nurse.

A stately white-tailed doe poses in a thicket near the Atchafalaya Basin in Louisiana. Does usually become sexually mature at the age of two, when they leave their mothers and go off to mate and have young of their own. Born after a gestation period of between 196 and 210 days, fawns are able to walk almost immediately after birth.

Credits

Cover—L. Rue, Jr., Bruce Coleman, Inc. 1—J. Shaw, B.C., Inc. 5—Wolfgang Bayer. 6–7—T. Dickinson, Photo Researchers, Inc. 9—Z. Leszczynski, Animals Animals. 10–11—R. Grogan, P.R., Inc. 12–13—W. Curtsinger, P.R., Inc. 14–15—Peter B. Kaplan. 16–17—Entheos. 19—J. Wright, B.C., Inc. 20—Wolfgang Bayer. 21 (top)—D. Guravich, P.R., Inc.; (bottom) Jen & Des Bartlett, B.C., Inc. 22—Ogden Tanner. 23 (top)—L. Rue, Jr., B.C., Inc.; (bottom) W. Curtsinger, P.R., Inc. 24—Wolfgang Bayer. 25—Jen & Des Bartlett, B.C., Inc. 26–27—W. Curtsinger, Rapho, P.R., Inc. 28–31—Wolfgang Bayer. 33—G. Scott, P.R., Inc. 34—Jen & Des Bartlett, B.C., Inc. 35—C. Lockwood, Animals Animals. 36–37—L.L. Rue, P.R., Inc. 37—Tom Brakefield. 40 (top)—G. Schaller, B.C., Inc.; (bottom) D. Norris, P.R., Inc. 41—Tom Brakefield. 42—S. Dalton, P.R., Inc. 43 (top)—Stouffer Productions, Photo Researchers, Inc.; (bottom, left)—Stouffer Productions, Animals Animals; (bottom, right)—S. Dalton, P.R., Inc. 44 (top)—B. Wilson, P.R., Inc.; (bottom) C. Lockwood, P.R., Inc. 45—T. Dickinson, P.R., Inc. 46–47—R. Kinne, P.R., Inc. 48 (left)—E. Ciampi, P.R., Inc.; 48–49—T. McHugh, Steinhart Aquarium, P.R., Inc.; 49—T. McHugh, Steinhart Aquarium, P.R., Inc. 54–55—M. Stouffer, B.C., Inc. 57–64—W.H. Amos, B.C., Inc. 65—D. Schwimmer, B.C., Inc. 66—Thase Daniel. 66–67—Entheos. 68–69—R. Carr, B.C., Inc. 71—J. Bishop, B.C., Inc. 72—J. Shaw, B.C., Inc. 73 (left)—D. Overcash, B.C., Inc.; (right) Tom Brakefield. 74 (top, left)—J. Carmichael, B.C., Inc.; (bottom, left) E. Degginger, B.C., Inc. 74–75—S. Krasemann, P.R., Inc. 78—Z. Leszczynski, Animals Animals. 79—Entheos. 80—Z. Leszczynski, Animals Animals. 81—C. Ott, P.R., Inc. 82 (top)—G. Meszaros, B.C., Inc.; 82–83—C. Sherman, P.R., Inc. 83 (top, left)—Tom Brakefield; (top, right) G. Meszaros, B.C., Inc. 88–89—Tom Brakefield. 90–91—L.L. Rue III, B.C., Inc. 93—J. Carmichael, Jr. B.C., Inc. 94—R. Davies, B.C., Inc. 95 (top)—W.H. Amos, B.C., Inc.; (bottom) Wolfgang Bayer. 96—R. Mendez, Animals Animals. 97—Z. Leszczynski, Animals Animals. 100–101—W. Amos, B.C., Inc.; 101 (left)—Oxford Scientific Films, B.C., Inc.; (right) W. Amos, B.C., Inc. 102 (top)—N. Fox-Davies, B.C., Inc.; (bottom) W. Amos, B.C., Inc. 103 (top)—H.N. Darrow, B.C., Inc.; (bottom) Dr. J.A. Cooke, B.C., Inc.; 104 (top)—W.H. Amos, B.C., Inc.; (bottom) R. Dunne, B.C., Inc. 105—E. Degginger, B.C., Inc. 106—A. Limont, B.C., Inc. 107 (left)—Thase Daniel; (right) J. Shaw, B.C., Inc. 108—J. Shaw, B.C., Inc. 109—Oxford Scientific Films, B.C., Inc. 111—R. Stocker, P.R., Inc. 112 (left, top)—R. Dunne, B.C., Inc.; (left, bottom) Kirtley-Perkins, P.R., Inc. 112–113—G. Zohm, B.C., Inc. 113—Thase Daniel. 114–115—Tom Brakefield. 115—D. Overcash, B.C., Inc. 116 (left)—Z. Leszczynski, Animals Animals; (top, right) J. Hall, Animals Animals; (bottom, right) Thase Daniel. 117—C. Lockwood, B.C., Inc. 119—W. Fraser, B.C., Inc. 120—Peter B. Kaplan. 120–121—W. Ruth, B.C., Inc., 121—Peter B. Kaplan. 122 (left)—Stouffer Productions, Animals Animals; 122–123—C. Lockwood, Animals Animals. 124–125—Thase Daniel.

Photographs on endpapers are used courtesy of Time-Life Picture Agency, Russ Kinne and Stephen Dalton of Photo Researchers, Inc., and Nina Leen.

Film sequence on page 8 is from "All-American Raccoon," a program in the Time-Life Television series *Wild, Wild World of Animals*.

ILLUSTRATIONS on pages 38–39 and 84–87 are by John Groth, the frontispiece illustration on page 50 is courtesy of Bettmann Archive, the painting on pages 52–53 is courtesy of The New York Public Library. The illustration on pages 76–77 is by André Durenceau, and the woodcut on pages 98–99 is by Mark Kseniak.

Bibliography

NOTE: Asterisk at the left means that a paperback volume is also listed in *Books in Print*.

Amos, William H., *The Life of the Pond*. McGraw-Hill, 1967.

——"Teeming Life of a Pond." *National Geographic*, August 1970, p. 274.

Barker, Will, *Familiar Reptiles and Amphibians of America*. Harper & Row, 1964.

Bartlett, Des and Jen, "Beavers." *National Geographic*, May 1974, p. 716.

Buchsbaum, Ralph, *Animals Without Backbones*. University of Chicago, Press 1948.

Buyukmihci, Hope S., *Hour of the Beaver*. Rand McNally, 1971.

Cahalane, Victor H., *Mammals of North America*. Macmillan, 1966.

*Carrighar, Sally, *One Day at Teton Marsh*. Alfred A. Knopf, 1947.

Carrington, Richard, and the editors of Time-Life Books, *The Mammals*. Time-Life Books, 1963.

Clausen, Lucy W., *Insect Fact and Folklore*. Macmillan, 1954.

Cochran, Doris, *Living Amphibians of the World*. Doubleday, 1961.

Collins, Henry Hill, *Complete Guide of American Wildlife*. Harper & Row, 1959.

*Conant, Roger, *A Field Guide to Reptiles and Amphibians of Eastern and Central North America*. Houghton Mifflin, 1975.

Eddy, Samuel, *How to Know the Freshwater Fishes*. Wm. C. Brown, 1969.

Ernst, Carl Y., and Barbour, Roger W., *Turtles of the United States*. University of Kentucky Press, 1972.

Farb, Peter, and the editors of Time-Life Books, *The Insects*. Time-Life Books, 1962.

Grzimek, Bernhard, *Grzimek's Animal Life Encyclopedia*, Vols. 1–7, 10–13. Van Nostrand Reinhold, 1972–1975.

Kirkland, Wallace, *The Lure of the Pond*. Henry Regnery, 1969.

Klots, Elsie B., *The New Field Book of Freshwater Life*. Putnam, 1966.

Milne, Lorus and Margery, *Invertebrates of North America*. Doubleday, 1972.

Neary, John, *Insects and Spiders*. Time-Life Films, 1977.

Niering, William A., *The Life of the Marsh*. McGraw-Hill, 1966.

North, Sterling, *Raccoons are the Brightest People*. Dutton, 1967.

Oulahan, Richard, *Reptiles and Amphibians*. Time-Life Films, 1976.

Park, Ed, *The World of the Otter*. J. B. Lippincott, 1971.

Pennak, Robert W., *Freshwater Invertebrates of the United States*. Ronald Press, 1953.

Platt, Rutherford, *Wilderness: The Discovery of a Continent of Wonder*. Dodd, Mead, 1961.

Redford, Polly, *Raccoons and Eagles*. Dutton, 1965.

*Reid, George K., *Pond Life*. Golden Press, 1967.

Russell, Franklin, *Watchers at the Pond*. Alfred A. Knopf, 1961.

*Thoreau, Henry David, *Walden*. Peter Pauper, 1966.

Usinger, Robert L., *The Life of Rivers and Streams*. McGraw-Hill, 1967.

Walden, Howard T., *Familiar Freshwater Fishes of America*. Harper & Row, 1964.

Walker, Ernest P., et al., *Mammals of the World*. Johns Hopkins, 1975.

*Zim, Herbert, and Cottam, Clarence, *Insects*. Golden Press, 1957.

Index